# ENHANCING DATA MANAGEMENT THROUGH THE STATISTICAL DATA AND METADATA EXCHANGE STANDARD

## A SPECIAL SUPPLEMENT OF THE KEY INDICATORS FOR ASIA AND THE PACIFIC 2024

AUGUST 2024

ASIAN DEVELOPMENT BANK

ADB

© 2024 Asian Development Bank
6 ADB Avenue, Mandaluyong City, 1550 Metro Manila, Philippines
Tel +63 2 8632 4444; Fax +63 2 8636 2444
www.adb.org

Some rights reserved. Published in 2024.

ISBN 978-92-9270-815-3 (print); 978-92-9270-816-0 (PDF); 978-92-9270-817-7 (ebook)
Publication Stock No. FLS240373-2
DOI: http://dx.doi.org/10.22617/FLS240373-2

The views expressed in this publication are those of the authors and do not necessarily reflect the views and policies of the Asian Development Bank (ADB) or its Board of Governors or the governments they represent.

ADB does not guarantee the accuracy of the data included in this publication and accepts no responsibility for any consequence of their use. The mention of specific companies or products of manufacturers does not imply that they are endorsed or recommended by ADB in preference to others of a similar nature that are not mentioned.

By making any designation of or reference to a particular territory or geographic area in this document, ADB does not intend to make any judgments as to the legal or other status of any territory or area.

Please contact pubsmarketing@adb.org if you have questions or comments with respect to content, or if you wish to obtain copyright permission for your intended use that does not fall within these terms, or for permission to use the ADB logo.

Corrigenda to ADB publications may be found at http://www.adb.org/publications/corrigenda.

Note:
In this publication, "$" refers to United States dollars.
ADB recognizes the "Kingdom of Bhutan" as Bhutan, the "Republic of Maldives" as Maldives, the "Independent State of Samoa" as Samoa, and the "Kingdom of Thailand" as Thailand.

Cover design by Claudette Rodrigo.

# CONTENTS

# TABLES, FIGURES, AND BOXES

## BOXES

# FOREWORD

In today's data-driven world, the ability to efficiently analyze, exchange, and disseminate statistical information is paramount for informed decision-making and effective policy formulation. From government agencies to private enterprises, organizations are overwhelmed with vast amounts of data collected from various sources. This abundance of data presents both opportunities and challenges that may need robust mechanisms to manage information effectively.

The wealth of available data and the entry of new data producers are combining to place pressure on official statisticians. To remain relevant, producers of official statistics are increasingly required to innovate and respond to change, without sacrificing data quality or statistical accuracy. For statisticians, innovation must consider more than just methods of data collection and handling: it must also encompass how organizations and the broader statistical system are functioning.

In response, the Statistical Data and Metadata eXchange (SDMX) standard has emerged as a means to streamline data activities and facilitate interoperability across the world. Developed and sponsored by eight major international organizations, SDMX provides a comprehensive framework for structuring, collecting, producing, exchanging, and managing statistical data and metadata. By doing so, SDMX enables seamless integration and sharing of data across different systems and domains.

The Asian Development Bank (ADB), through the Data Division of the Economic Research and Development Impact Department, has positioned itself to play a key role in the adoption of SDMX in its developing member countries (DMCs) across Asia and the Pacific. ADB is committed to supporting the ongoing development of the national statistical systems within DMCs, and has consistently advocated for high-quality statistical information using modern technology, innovative data, advanced methods, and best practices. In relation to SDMX, we have enhanced the capacity of DMCs and conducted SDMX training for staff from national statistics offices and other data-producing agencies. This includes the creation (in collaboration with development partners) of e-learning courses on SDMX foundations and key SDMX tools. More than 600 participants from over 65 countries around the world have successfully completed these online training courses.

This year's special supplement to *Key Indicators for Asia and the Pacific* highlights the advantages of adopting the SDMX standard, particularly in developing economies. It presents the results of an ADB technical assistance project that was launched in 2018 to promote the adoption of the SDMX standard in our region. Under the project, and in collaboration with development partners, ADB has supported the National Statistical Office of Thailand to implement SDMX within their decentralized reporting system. The office's staff were trained in SDMX concepts, data modeling using globally defined data structures, and customizing data structures according to national requirements.

This special supplement is a product of close collaboration among a team of experts drawn from a variety of disciplines. The supplement team was led by Stefan Schipper, under the overall guidance of Elaine S. Tan. Stefan was joined by Brian Buffett and Jeffrey Napoles as coauthors of the report, while significant contributions in research and technical support were made by Pamela Lapitan and Thomas John Ballatore. The supplement team extends its appreciation to the participating national statistics offices and other data-producing agencies, with special recognition to the National Statistical Office of Thailand, the Samoa Bureau of Statistics, and the National Statistics Bureau of Bhutan. Their invaluable contributions have been instrumental in driving the implementation of SDMX and strengthening the capabilities of statistical systems in other DMCs. The team would also like to acknowledge Denis Grofils from the Pacific Community for his invaluable insights and experiences regarding SDMX implementation, as well as Dayyan Shayani from the United Nations Economic and Social Commission for Asia and the Pacific for his input on SDMX support in Maldives. Additionally, the team is grateful to Abdulla Gozalov, Hernan Hernandez Martinez, and Markie Muryawan of the United Nations Statistics Division, as well as Pinar Ucar from the United Nations Statistical Institute for Asia and the Pacific, for their invaluable contributions and support for the SDMX e-Learning courses.

We hope this publication can play a role in promoting the adoption of SDMX throughout Asia and the Pacific, serving as a catalyst for standardization and modernization in the management and exchange of statistical data and metadata—leading ultimately to more accurate policy development and more equitable allocation of development resources.

**Albert F. Park**
Chief Economist and Director General
Economic Research and Development Impact Department
Asian Development Bank

# ABBREVIATIONS

| | |
|---|---|
| ADB | Asian Development Bank |
| API | application programming interface |
| COGs | content-oriented guidelines |
| CSV | comma-separated values |
| DMC | developing member country |
| DSD | data structure definition |
| e-GDDS | Enhanced General Data Dissemination System |
| Eurostat | Statistical Office of the European Union |
| IMF | International Monetary Fund |
| ISO | International Organization for Standardization |
| JSON | JavaScript Object Notation |
| NSO | national statistics office |
| NSS | national statistical system |
| SDDS | Special Data Dissemination Standard |
| SDG | Sustainable Development Goal |
| SDMX | Statistical Data and Metadata eXchange |
| SIS-CC | Statistical Information System Collaboration Community |
| SPC | The Pacific Community |
| TNSO | National Statistical Office of Thailand |
| UNSC | United Nations Statistical Commission |
| VTL | Validation and Transformation Language |
| XML | Extensible Markup Language |

Note: In this report, the term "economy" is widely used to represent "country" or "nation". Any use of the terms "country", "nation", or "national" is not intended to make any judgment as to the legal or other status of any territory or area.

# HIGHLIGHTS

- **Statistical Data and Metadata eXchange (SDMX)** is an international standard for improving the efficiency, quality, and accessibility of statistical data. The main objectives of SDMX are to standardize and modernize the collection, processing, and exchange of statistical data and metadata. These factors can contribute significantly to more informed decision-making and policy formulation at both national and international levels.

- **SDMX is a global initiative sponsored by eight major international organizations:** the Bank for International Settlements, the European Central Bank, the International Labour Organization, the International Monetary Fund, the Organisation for Economic Co-operation and Development, the Statistical Office of the European Union (Eurostat), the United Nations, and the World Bank. These sponsor organizations have instituted and maintained common SDMX technical and statistical standards and guidelines, and they continue to collaborate to enhance statistical data and metadata management through SDMX. Together, they are working to streamline data and metadata exchange processes, offer capacity development programs, provide information technology infrastructure for efficient data sharing, and explore the need for future evolvement of the standards.

- **SDMX is the preferred standard of the United Nations Statistical Commission (UNSC) for the exchange of data and metadata.** In 2008, the UNSC endorsed SDMX as the preferred standard for the exchange and sharing of both data and metadata, with members of the UNSC collectively agreeing that SDMX provides a robust and effective solution for standardizing practices related to data and metadata exchange.

- **SDMX is an International Organization for Standardization (ISO) standard** (ISO 17369:2013). It is specifically designed to foster interoperable implementations within and between systems, and be applicable to any organization responsible for the collection, processing, and exchange of statistical data and associated metadata.

- **SDMX supports modernization of statistical data management practices** by offering standardized and interoperable solutions, adopting new technologies, facilitating efficient tool development, encouraging worldwide collaboration, and aligning with contemporary data-sharing initiatives and global development goals.

- **SDMX offers numerous benefits for the management of statistical data**, including enhanced efficiency, interoperability, and standardization. By providing a common framework for data and metadata exchange and sharing, the SDMX standard facilitates seamless communication between different systems and organizations. This standardized approach also improves data quality and timeliness, reduces duplication of effort, and enhances the comparability of statistical information across various domains.

- **Key issues and challenges remain within national statistical systems.** Issues and challenges within national statistical systems include obstacles faced in collecting, integrating, and disseminating statistical data. These include ensuring data quality, addressing resource limitations, navigating legal and ethical considerations, and adopting modern technology.

- **Free, high-quality, open-source software tools are available to facilitate SDMX adoption.** There is an array of freely accessible tools available to aid in the implementation of SDMX. These tools serve different purposes within the SDMX framework, such as data modeling, data collection, data conversion, data validation, metadata management, and data dissemination.

- **Economies of Asia and the Pacific emphasize a growing need for SDMX.** Survey results from Asian Development Bank (ADB) members across Asia and the Pacific highlight a notable emphasis on adopting and utilizing the SDMX standard, especially for enhancing data and metadata dissemination, as well as improving metadata management and standardizing statistical business processes.

- **SDMX Implementation in Thailand demonstrates the benefits of data harmonization.** The National Statistical Office of Thailand (TNSO) has undertaken a series of initiatives to implement SDMX within their decentralized reporting system. The TNSO's journey began with the development of a data exchange system and progressed to more robust SDMX-based infrastructure, such as the TNSO Statistical Sharing Hub.

- **ADB supports upskilling via SDMX e-learning courses.** In collaboration with three development partners, ADB has developed two vital e-learning courses on SDMX. The **SDMX foundation course** provides an introduction to SDMX and its most important components, such as the information model, content-oriented guidelines, and available SDMX infrastructure and information technology. The **SDMX tools course** is a follow-up to the foundational course and looks in depth at three tools: the SDMX Constructor, the Fusion Metadata Registry, and the SDMX Converter.

- **Both e-learning courses have achieved high completion rates and positive ratings.** The SDMX foundations course was first conducted from 28 March to 15 April 2022. Its completion rate was 91.9% and its approval rating (graded as *good* or *excellent* by participants) was 94.9%. The SDMX tools course was first conducted from 15 November to 15 December 2023. Given the significantly more demanding nature of the content, the completion rate was still exceptionally high for an online course, at 39.4%. The approval rating was also very high, with 94% of participants grading it as good or excellent.

Statistical Data and Metadata eXchange (SDMX) is an international standard established to improve the efficiency, quality, and accessibility of statistical data. The SDMX framework provides guidance on standardizing and modernizing the collection, processing, and exchange of statistical data and metadata (SDMX 2023).

SDMX is driven by the need for consistency in interpreting and presenting statistical information. The standard provides a common language and a set of rules for describing, exchanging, and validating data and metadata across different statistical domains, such as national accounts, balance of payments, and foreign direct investments.

As an illustrative example, when datasets are released only in spreadsheet or digital file format, and are not stored in a centralized database, users may struggle to find the specific data they need. Moreover, reporting to international organizations also becomes challenging since national statistics offices are required to convert the data into acceptable formats, increasing workload and potential inconsistencies. SDMX addresses these challenges by using standardized structures that ensure consistency across datasets and eliminate the need for multiple format conversions. Additionally, the standard allows for a centralized repository where all data are published, streamlining access for users.

This commitment to standardized practices not only streamlines the data exchange process but also enhances the overall utility and accessibility of statistical information on a global scale.

Importantly, the SDMX standard enables the automation and integration of data flows and processes, reducing the resource and cost burden of data reporting and enhancing the timeliness and quality of the data collected. It also supports the interoperability and accessibility of data, which allows statisticians and relevant stakeholders to easily access, analyze, and apply the data for their own specific purposes.

## 1.1    Origins and Evolution of Statistical Data and Metadata eXchange

The SDMX global initiative is sponsored by eight major international organizations: the Bank for International Settlements, the European Central Bank, the International Labour Organization, the International Monetary Fund, the Organisation for Economic Co-operation and Development, the Statistical Office of the European Union (Eurostat), the United Nations, and the World Bank. As well as collaborating on the SDMX framework itself, the sponsor organizations have instituted and maintained common SDMX technical and statistical standards and guidelines.

Since its establishment in 2001, the SDMX initiative has made remarkable progress, with the collective efforts of the eight sponsor organizations playing a pivotal role in its development.

In 2008, the United Nations Statistical Commission (UNSC) officially acknowledged and endorsed SDMX as the preferred standard for the exchange and sharing of both data and metadata (UNSC 2008). This endorsement implies a collective agreement among the members of the UNSC that SDMX provides a robust and effective solution for standardizing practices related to data and metadata exchange. As a preferred standard, SDMX is encouraged for adoption across United Nations agencies and affiliated organizations to ensure a harmonized approach to statistical data management and dissemination (UNSC 2008).

In 2013, the International Organization for Standardization (ISO) confirmed SDMX as an official international standard (ISO 17369:2013). Under the ISO definition, SDMX is specifically applied to foster interoperable implementations within and between systems, and is applicable to any organization with a mandate to manage the reporting, exchange, and dissemination of statistical data and associated metadata. The information model of ISO 17369:2013 was established to support statistics in alignment with the practices of national governments and supranational statistical organizations. However, use of the information model extends beyond these organizations, proving applicable to diverse organizational contexts involving the management of statistical data and metadata (ISO 2023).

Figure 1 provides an overview of the evolution of SDMX, showcasing key milestones achieved through the collaborative efforts of SDMX community practitioners from both the public and private sectors.

## 1.2 Structure, Applications, and Benefits of Statistical Data and Metadata eXchange

There are three main components of the SDMX standard: the information model, content-oriented guidelines, and the technical standard.

The information model defines the conceptual framework for structuring statistical data and metadata, ensuring a standardized understanding of information.

The content-oriented guidelines provide procedures and best practices for creating and managing the content of statistical data and metadata within the SDMX framework. For instance, the SDMX list of economy codes can be used universally because each economy is assigned a unique, language-independent code, e.g., "FR" for France, "JP" for Japan, and "US" for the United States.  These codes remain consistent regardless of the language used in the dataset. So, whether the dataset is presented in French, Japanese, English, or any other language, the economy codes in the SDMX codelist remain standardized.

 The technical standard refers to the software architecture definitions that enable the production of computing tools and online data services in accordance with the information model and the content-oriented guidelines. These definitions work together to establish a common language and framework for the exchange of statistical information (SDMX 2020a).

## Figure 1: Key Milestones for Statistical Data and Metadata eXchange

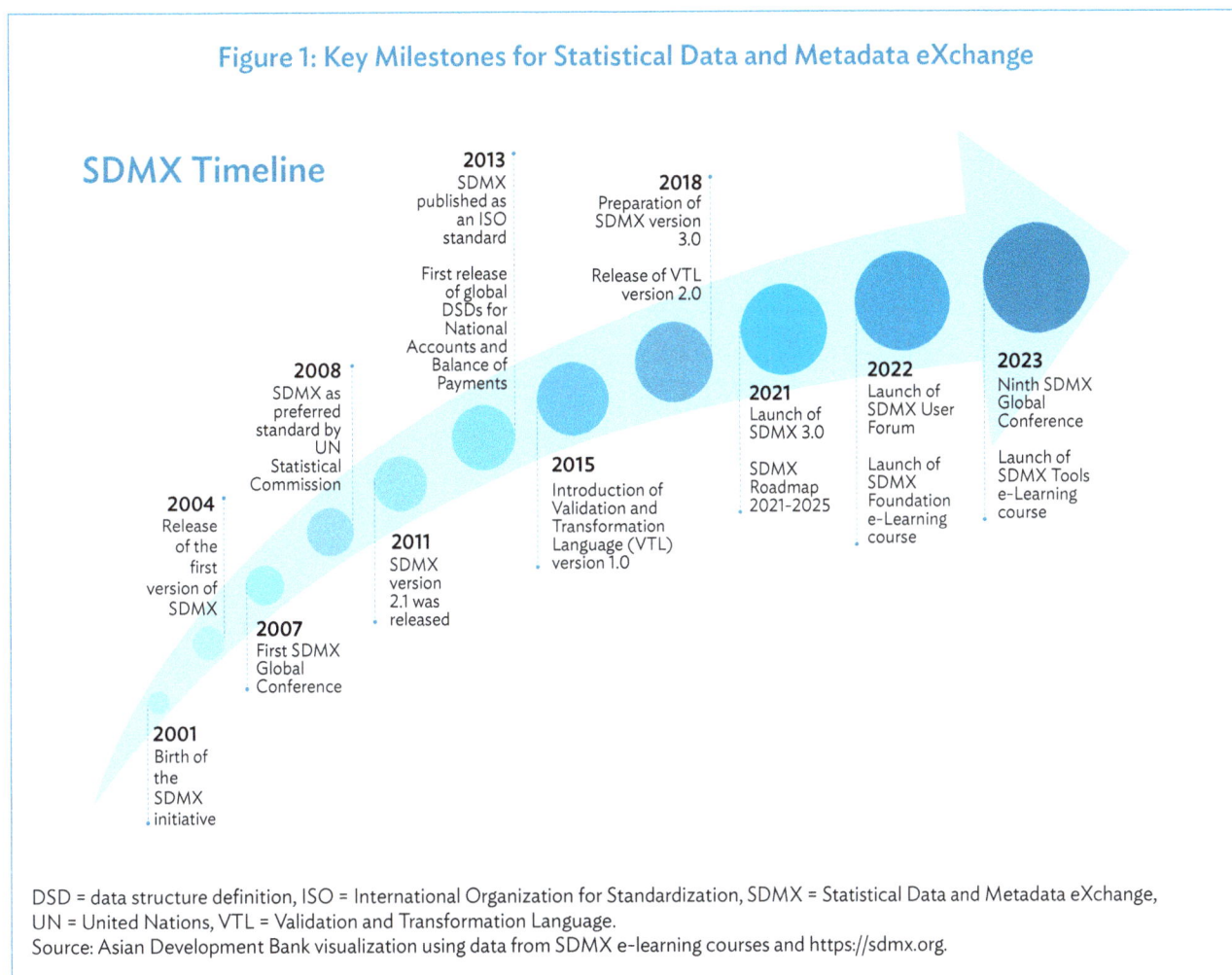

### SDMX Timeline

**2001**
Birth of the SDMX initiative

**2004**
Release of the first version of SDMX

**2007**
First SDMX Global Conference

**2008**
SDMX as preferred standard by UN Statistical Commission

**2011**
SDMX version 2.1 was released

**2013**
SDMX published as an ISO standard

First release of global DSDs for National Accounts and Balance of Payments

**2015**
Introduction of Validation and Transformation Language (VTL) version 1.0

**2018**
Preparation of SDMX version 3.0

Release of VTL version 2.0

**2021**
Launch of SDMX 3.0

SDMX Roadmap 2021-2025

**2022**
Launch of SDMX User Forum

Launch of SDMX Foundation e-Learning course

**2023**
Ninth SDMX Global Conference

Launch of SDMX Tools e-Learning course

DSD = data structure definition, ISO = International Organization for Standardization, SDMX = Statistical Data and Metadata eXchange, UN = United Nations, VTL = Validation and Transformation Language.
Source: Asian Development Bank visualization using data from SDMX e-learning courses and https://sdmx.org.

One of the standout features of the technical standard for SDMX is the Application Programming Interface (API), which empowers users to efficiently retrieve structured datasets and seamlessly integrate them into statistical systems.

Various international organizations have collaborated to produce and share SDMX global data structures that can be used for different statistical domains and purposes. Global data structures exist for national accounts, balance of payments, consumer price indices, labor force statistics, Sustainable Development Goal (SDG) indicators, and more. These data structures are based on common concepts and classifications widely accepted and used by the statistics community. For instance, the SDG data structure includes disaggregation factors such as sex, age, and urbanization as integral components that can be recognized globally.

In the compilation and dissemination of data for the SDG indicators, the means of officially measuring progress toward SDG targets, the SDMX standard plays a key role. A predetermined format and structure for SDG data has been defined in the SDMX standard, which assists in harmonizing, validating, aggregating, and disseminating a wide range of SDG datasets across different sources and platforms.

The use of SDMX global data structures benefits data-producing agencies by making them part of a larger statistical community. It also benefits data users by providing them with high-quality and consistent datasets that observe common definitions and formats, facilitating clearer interpretations and more accurate application of statistical information. Importantly, using SDMX global data structures also benefits both data producers and users through reduced costs for data collection, processing, analysis, and dissemination. SDMX improves the efficiency and quality of data generation and dissemination by reducing manual work, errors, and inconsistencies and by ensuring compliance with international standards and definitions.

## 1.3  Ongoing Adoption and Uptake of Statistical Data and Metadata eXchange

The eight SDMX sponsor organizations have been collaborating to conduct capacity development programs, training courses, and workshops for national agencies involved in the production and exchange of data, aiding these agencies in the modernization of their statistical systems. In addition, the sponsor organizations have leveraged the SDMX technical standard to develop numerous free, open-source software tools to promote more efficient definition, collection, production, exchange, and management of statistical data and metadata (SDMX 2023). As the SDMX standard has gained traction, other entities such as national statistics offices, central banks, and private firms have developed open-source and commercial SDMX tools and connectors.

# 2 CORE USES OF STATISTICAL DATA AND METADATA EXCHANGE

Statistical Data and Metadata eXchange (SDMX) plays a crucial role in standardizing the exchange of information across different stages of the statistical process, including data collection, production, reporting, and dissemination. Use of the standard promotes interoperability and consistency in the way data is represented, making it more efficient for organizations to share and use readily understood statistical information at national and international levels.

## 2.1 Data Collection

Data collection refers to the process of gathering statistical source information by various means, such as surveys, censuses, polls, or administrative records.

SDMX standardizes the collection of data by providing a common framework for defining data structures, codes, and classifications. It ensures that collected data conform to a specified format, making it easier to aggregate, compare, and analyze information from diverse sources. For example, under the SDMX standard, geographic locations or individual economies are each identified by a unique and universally recognized code, such as "PH" for the Philippines and "TH" for Thailand.

## 2.2 Data Validation and Statistics Production

Statistics production is where the complex and highly specialized tasks of validating and aggregating data take place, along with the analysis of high-quality statistics.

SDMX supports statistics production with the introduction of metadata-driven processes and tools. In this context, a metadata-driven process is a systematic approach to modeling, structuring, and organizing the information about data. This benefit increases even further if validation, transformation, and aggregation processes are driven by metadata that identify and describe data.

## 2.3 Data Reporting

Data reporting involves the process of submitting statistical data to a central authority or repository. This could be a national statistics office, an international organization, or any other entity responsible for collecting and managing statistical information.

SDMX provides a standardized format for structuring and transmitting data, ensuring that data reported by different entities are compatible and can be easily integrated. The SDMX standard specifies how data should be organized. Figure 2 illustrates an SDMX workflow from a national statistics office to an international organization for reporting Sustainable Development Goal (SDG) data.

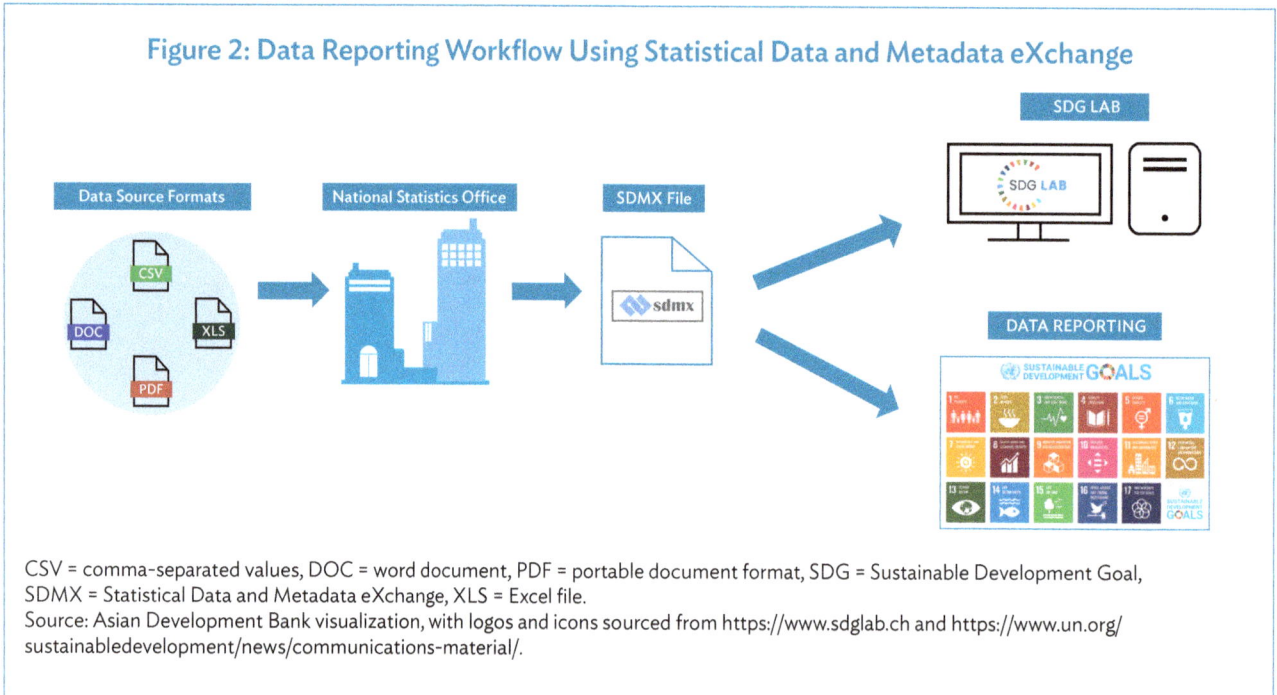

**Figure 2: Data Reporting Workflow Using Statistical Data and Metadata eXchange**

CSV = comma-separated values, DOC = word document, PDF = portable document format, SDG = Sustainable Development Goal, SDMX = Statistical Data and Metadata eXchange, XLS = Excel file.
Source: Asian Development Bank visualization, with logos and icons sourced from https://www.sdglab.ch and https://www.un.org/sustainabledevelopment/news/communications-material/.

## 2.4 Data Dissemination

Data dissemination involves making statistical information available to users, including the general public, policymakers, researchers, and other stakeholders.

SDMX supports data dissemination by defining standardized structures for the metadata used to describe the content and quality of statistical data, such as their sources, methodologies, and quality indicators. This metadata helps users understand the context and characteristics of the data being disseminated. Additionally, SDMX enables the publication of data in a format that is easily accessible and understandable.

## 2.5 The Role of Application Programming in Data Dissemination

Utilizing an application programming interface (API) is an effective means of disseminating data. An SDMX-API offers a standardized framework of protocols and methods made for efficiently requesting statistical data. By using an SDMX-API, data retrieval processes can be streamlined through automation, empowering applications and systems to interact with SDMX-enabled data sources. This capability facilitates scheduled, repeatable, and automated tasks for retrieving data.

SDMX-API not only simplifies the process but also enables direct access to statistical databases and repositories. This supports real-time data updates, allowing applications to retrieve the most recent information directly from the data source. This is particularly important where up-to-date data are critical, such as data used for economic indicators, financial modeling, or other time-sensitive statistical activities. In essence, an SDMX-API serves as a powerful tool to enhance the efficiency and timeliness of data dissemination and retrieval processes.

## 2.6 Leading Warehouses in High-Tech Data Retrieval

Several data warehouses around the world are equipped with an SDMX-API for data retrieval.[1] The Key Indicators Database - ADB's central statistical database housing economic, social, and environmental indicators from across Asia and the Pacific - has integrated the use of SDMX-API for efficient data retrieval and dissemination.

The United Nations Children's Fund (UNICEF) Indicator Data Warehouse has implemented SDMX Fusion Metadata Registry, which serves as a structural metadata repository for a wide range of indicators on health, education, nutrition, child protection, and more (Figure 3). The data warehouse incorporates a user-friendly interface using an SDMX-compliant representational state transfer (REST) API. It offers an intuitive interface for users to query, download, and export desired data and metadata by either selecting from the options or entering specific terms in the search field.[2]

**Figure 3: Data Retrieval via UNICEF Indicator Data Warehouse**

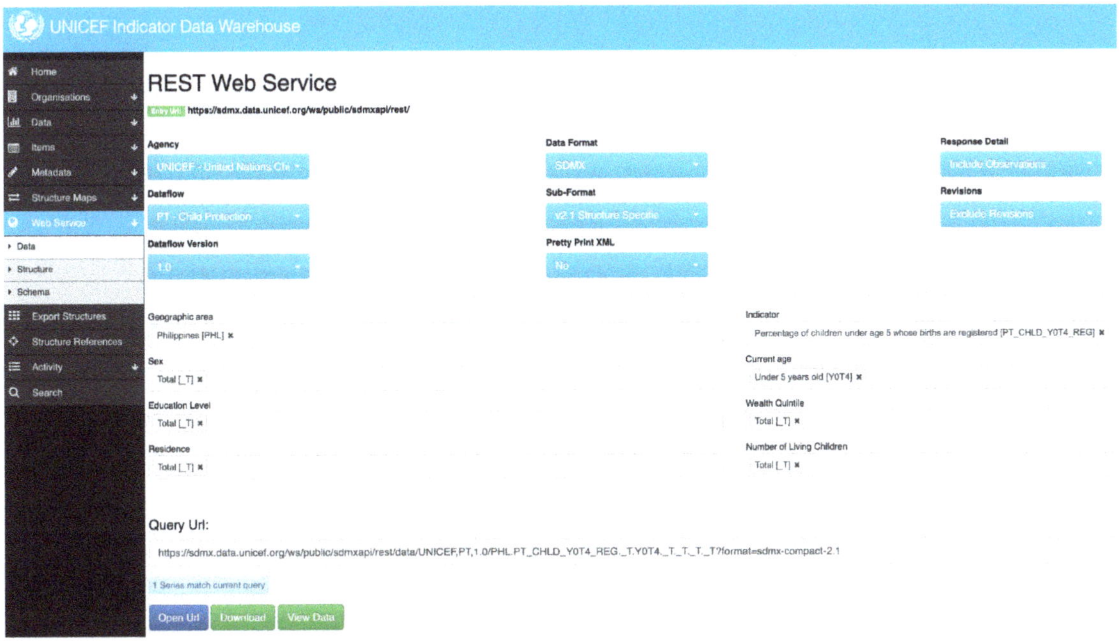

REST = representational state transfer, UNICEF = United Nations Children's Fund.
Source: Asian Development Bank screenshot taken from https://sdmx.data.unicef.org/webservice/data.html.

---

[1]  Access the Key Indicators Database via https://kidb.adb.org.
[2]  Access the UNICEF REST Web Service via  https://sdmx.data.unicef.org/webservice/data.html.

The Statistics Sharing Hub maintained by the National Statistical Office of Thailand is a comprehensive data portal containing a diverse array of international and national indicators (Figure 4). This hub uses the .Stat Suite data dissemination platform developed by the Statistical Information System Collaboration Community (SIS-CC).

Figure 4: Main Portal of Thailand's Statistics Sharing Hub

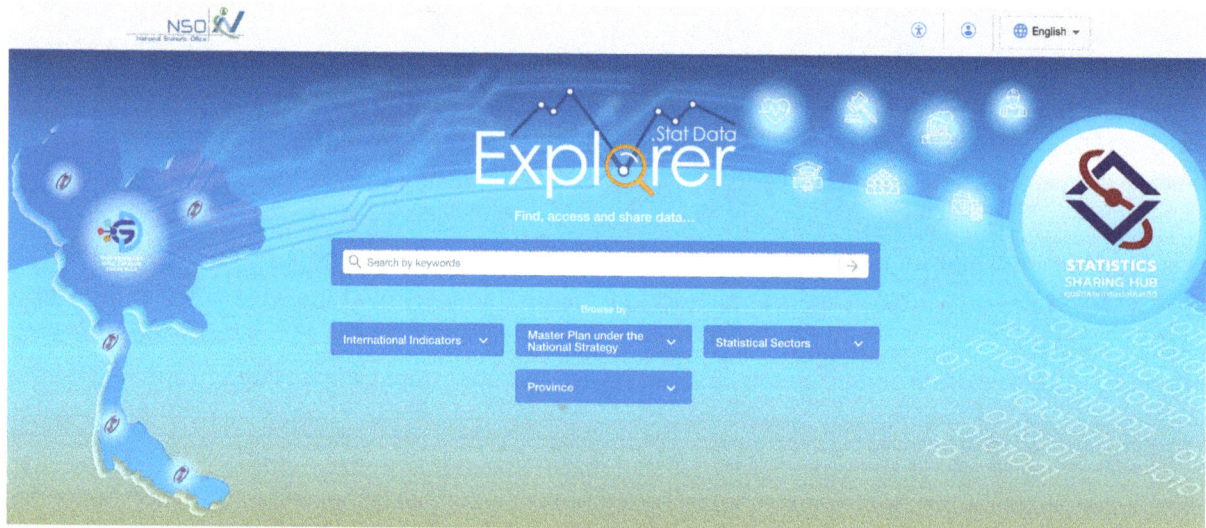

Source: Asian Development Bank screenshot taken from https://stathub.nso.go.th/?lc=en&pg=0 .

In addition to making data available via a rich tabular presentation tool, one of the key features of the hub is the availability of an API, enabling data portal users to access their desired data in a variety of ways (Figure 5).[3]

---

[3] Access the National Statistical Office of Thailand Statistics Sharing Hub via https://stathub.nso.go.th/?lc=en&pg=0.

## Figure 5: Application Programming Interface of Thailand's Statistics Sharing Hub

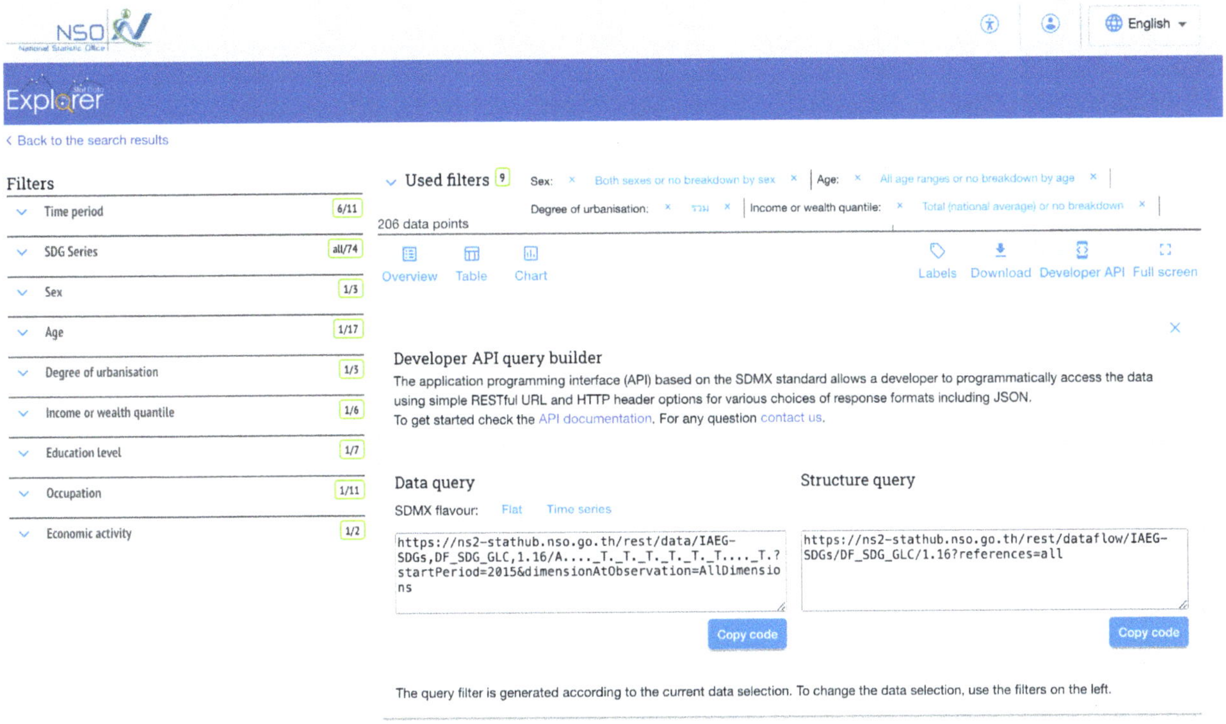

NSO
National Statistic Office

Explorer

< Back to the search results

**Filters**

| | | |
|---|---|---|
| ∨ | Time period | 6/11 |
| ∨ | SDG Series | all/74 |
| ∨ | Sex | 1/3 |
| ∨ | Age | 1/17 |
| ∨ | Degree of urbanisation | 1/3 |
| ∨ | Income or wealth quantile | 1/6 |
| ∨ | Education level | 1/7 |
| ∨ | Occupation | 1/11 |
| ∨ | Economic activity | 1/2 |

∨ **Used filters** 9    Sex:  ✕  Both sexes or no breakdown by sex  ✕  | Age:  ✕  All age ranges or no breakdown by age  ✕  |

206 data points    Degree of urbanisation:  ✕  รวม  ✕  | Income or wealth quantile:  ✕  Total (national average) or no breakdown  ✕  |

Overview    Table    Chart                    Labels    Download    Developer API    Full screen

✕

**Developer API query builder**

The application programming interface (API) based on the SDMX standard allows a developer to programmatically access the data using simple RESTful URL and HTTP header options for various choices of response formats including JSON.
To get started check the API documentation. For any question contact us.

**Data query**                                          **Structure query**

SDMX flavour:    Flat    Time series

```
https://ns2-stathub.nso.go.th/rest/data/IAEG-
SDGs,DF_SDG_GLC,1.16/A...._T._T._T._T._T._T...._T.?
startPeriod=2015&dimensionAtObservation=AllDimensio
ns
```

```
https://ns2-stathub.nso.go.th/rest/dataflow/IAEG-
SDGs/DF_SDG_GLC/1.16?references=all
```

Copy code                                          Copy code

The query filter is generated according to the current data selection. To change the data selection, use the filters on the left.

Source: Asian Development Bank screenshot taken from https://stathub.nso.go.th/vis?lc=en&pg=0&fs[0]=International%20
Indicators%2C0%7CSustainable%20Development%20Goals%23SDG%23&fc=International%20Indicators&bp=true&snb=1&vw=o-
v&df[ds]=ds-stathub-release&df[id]=DF_SDG_GLC&df[ag]=IAEG-SDGs&df[vs]=1.16&pd=2015%2C&dq=A....._T._T._T._T._T._T...._T.&ly[
cl]=TIME_PERIOD&ly[rw]=SERIES&to[TIME_PERIOD]=false.

# 3 KEY BENEFITS OF STATISTICAL DATA AND METADATA EXCHANGE

In this era of information abundance, understanding and harnessing the benefits of Statistical Data and Metadata eXchange (SDMX) is crucial for organizations navigating the complexities of data management and for policymakers seeking to make informed decisions based on a foundation of standardized and harmonized statistical data. As the demand for timely, accurate, and harmonized data continues to grow, SDMX provides a robust framework that facilitates the efficient exchange, integration, and dissemination of statistical data.

*The Business Case for SDMX* delves into the multifaceted advantages that SDMX can bring, shaping a scenario whereby data becomes not just a resource but a powerful tool for informed decision-making and evidence-based policies (SDMX 2020a). Some of the key benefits outlined in the SDMX business case report are detailed below.

## 3.1 Improved Data Quality and Timeliness

Adoption of the SDMX standard facilitates the production of more timely and better-quality data. It does so by reducing manual efforts, automating checks and workflows, enhancing accessibility, and minimizing the risk of errors throughout the data-exchange process.

SDMX promotes timeliness by minimizing the need for manual conversion of data. The use of automated checks in SDMX leads to swift validation of data. Automated validation processes can quickly identify errors or inconsistencies, reducing the time needed for manual verification and validation. For example, when reporting on the percentage of women in Parliament, an automated check ensures that the value for the sex field is correctly set to "female" and prevents any erroneous entry of "male" or "other". Such automation results in validated data being available more quickly and lets users access this high-quality data in a shorter time span.

SDMX also enables automated processing of data, reducing the likelihood of human errors associated with manual handling. Automated workflows ensure consistency and accuracy in data processing, contributing to better data quality overall.

Furthermore, adoption of the SDMX standard triggers increased automation through the implementation of automated workflows for the exchange of statistics (SDMX 2020a).

Significantly, the Validation and Transformation Language (VTL) initiative for SDMX is intended to enhance data quality further. VTL provides standardized instructions for expressing the validation and transformation rules applied to statistical data, improving both accuracy and consistency in the datasets (SDMX 2020b). The development of tools supporting VTL is ongoing.

## 3.2  Enhanced Data Consistency and Comparability

SDMX's standardized approach to organizing and exchanging data and metadata enables interoperable functionality within and between systems dedicated to exchanging, reporting, and disseminating statistical information. The result is improved data consistency and comparability across different statistical applications and organizations.

The power behind SDMX lies in its information model, which allows for the development of processes and functions around the data model, rather than being constrained by specific data syntaxes or formats. The SDMX information model establishes a common language and structure for expressing statistical concepts to ensure consistency and uniformity in the representation of data.

SDMX supports a common terminology for describing statistical data, harmonizing concepts and codelists. This strong terminological foundation standardizes data description across various statistical domains. In section 5.2.1, cross-domain concepts are explained in more detail, demonstrating how codelists and data description can be applied in more than one statistical domain in a similar form. Harmonizing statistical data content and structure offers numerous advantages, including a shared language among implementers and users. This is achieved through the use of uniform codes, with names and descriptions that can be expressed in different languages. This approach saves time and resources through reduced mapping and data processing, wider availability of tools based on a commonly agreed format, and the existence of SDMX registries that facilitate reuse.

SDMX also enhances interpretability by using a standardized terminology to harmonize structural and reference metadata. This contributes to the development of a global statistical language and improves coherence through cross-domain concepts, shared codelists, harmonized statistical guidelines, and extensive reuse of SDMX objects across domains and agencies (SDMX 2020a).

The adoption of the SDMX standard therefore promotes collaboration, consistency, and comparability in data sharing, underpinning the success of global initiatives to advance statistical data.

## 3.3  Minimized Burden of Data Reporting

The core responsibilities of national statistics offices (NSOs) include collecting, processing, and disseminating statistical data. Each organization that requests data from, or provides data to, an NSO may have its own unique data format requirements. In these instances, the NSO may find itself having to adapt and transform the same dataset into multiple formats, (e.g., spreadsheet, text file, digital file) significantly increasing its workload and resource requirements. Adoption of the SDMX standard proves highly beneficial in resolving this scenario.

Because SDMX promotes the use of standardized data structures and exchange formats, organizations can align their reporting systems with these standards, reducing the need for maintaining multiple, diverse reporting systems. Standardization across organizations and data domains further minimizes the complexity associated with managing various reporting formats and exchange agreements. This streamlining effect contributes to a significant reduction in the overall data reporting burden.

SDMX ensures data accuracy through prevalidation, automating the reporting process, standardizing data structures, and enabling efficient data collection methods. These attributes of the standard enhance the overall efficiency of the data reporting workflow, minimize manual interventions, and lead to a more streamlined and cost-effective reporting process.

The SDMX standard allows for the prevalidation of data messages against predefined data structures, also known as  data structure definitions. This ensures that the data messages are aligned with the structure, reducing errors during data exchange and minimizing the need for manual intervention. As a result, data producers can check the accuracy and adherence of their dataset to the specified structure before sending it. By avoiding errors and ensuring compliance at the source, this prevalidation feature minimizes the need for multiple rounds of communication between data producers and consumers to rectify data issues. It streamlines the reporting process and reduces the back-and-forth exchanges that often occur during data validation (SDMX 2020a).

## 3.4  Availability of Free and Open-Source Tools

Most of the SDMX information technology (IT) tools developed by various organizations are free and open-source. SDMX follows an open-source approach, meaning that the standards and tools associated with SDMX are freely available for anyone to use. This fosters a large community of developers and users who contribute to the improvement and evolution of the SDMX ecosystem. This open-source collaboration not only reduces the cost of acquiring tools but also allows organizations to benefit from a wide range of SDMX solutions created and maintained by the community. Additionally, the open-source community shares expertise and best practices, providing valuable resources at no cost[4] (SDMX 2020a).

## 3.5  Affordable Implementation and Easier Access to Data

SDMX is a versatile and accessible framework for organizations involved in statistical data management. It provides flexibility for customization and promotes cost-efficiencies via access to freely available development tools and codes. These factors go a long way to removing barriers to implementation and data accessibility.

The flexibility of the SDMX toolkit approach, which provides a set of standardized tools and resources that organizations can adapt to their specific needs, allows flexibility in the SDMX implementation process. Organizations are able to tailor SDMX implementation based on their unique requirements, infrastructure, and data-production processes. SDMX implementation can have a wide range of applications. It can resolve specific tasks such as data reporting, as shown in the National Data Summary Page (Section 7.1), or it can be used in comprehensive solutions for the entire data life cycle, exemplified by the Statistical Sharing Hub developed and maintained by the National Statistical Office of Thailand (Section 7.2).

---

4    Consult sdmx.org (https://sdmx.org) and sdmx.io (https://sdmx.io/tools/ecosystem) for more information on available tools.

Meanwhile, the open-source tools and codes that support SDMX eliminate the need for organizations to invest in proprietary software, often necessary via licensing fees. These open-source solutions contribute to cost-efficiencies, making SDMX more accessible to a broader range of entities, especially those with limited budgets (SDMX 2020a).

## 3.6  Streamlined Integration and Improved Statistical Practices

SDMX is designed to be flexible and interoperable with modern technologies. This adaptability allows NSOs and other data-producing agencies to integrate SDMX with contemporary IT infrastructures, databases, and data-processing tools, promoting a more modern and efficient data ecosystem.

The adoption of SDMX also contributes to the modernization of statistical data management practices on a worldwide scale. Because the standard is globally endorsed and adopted, it encourages collaboration among international organizations and national economies and fosters the exchange of modern practices, methodologies, and technologies, creating a shared and regularly updated vision for statistical data management (SDMX 2020a).

## 3.7  Stability Through Strong Commitment and Governance

The SDMX initiative is sponsored by eight major international organizations. The SDMX sponsors, together with other development partners, collaborate closely with NSOs and other data-producing agencies worldwide. The involvement of these reputable entities lends credibility to SDMX as a global standard. Moreover, the commitment of the sponsor organizations spans more than 20 years, indicating a long-term dedication to the success and continuity of SDMX. This commitment ensures stability and reliability in the implementation and development of the standard (SDMX 2020a).

The SDMX initiative also benefits from a well-established governance model, as shown in Figure 6.

Key entities within the governance structure include the Sponsors Committee, which makes strategic decisions, and the Secretariat, which is responsible for the operational management of the initiative. This hierarchical structure allows for effective decision-making and coordination among the SDMX sponsor organizations. Meanwhile, the Technical Working Group and the Statistical Working Group operate with a proactive approach that ensures the SDMX standard evolves according to the changing needs of users and remains relevant and adaptable.

The initiative's governance model also offers a proven approach for managing the life cycle of a statistical domain's SDMX objects and versions. This structured approach ensures that SDMX evolves systematically and updates are managed efficiently, contributing to the long-term reliability and stability of the standard.

Figure 6: Governance Structure for the Statistical Data and Metadata eXchange Initiative

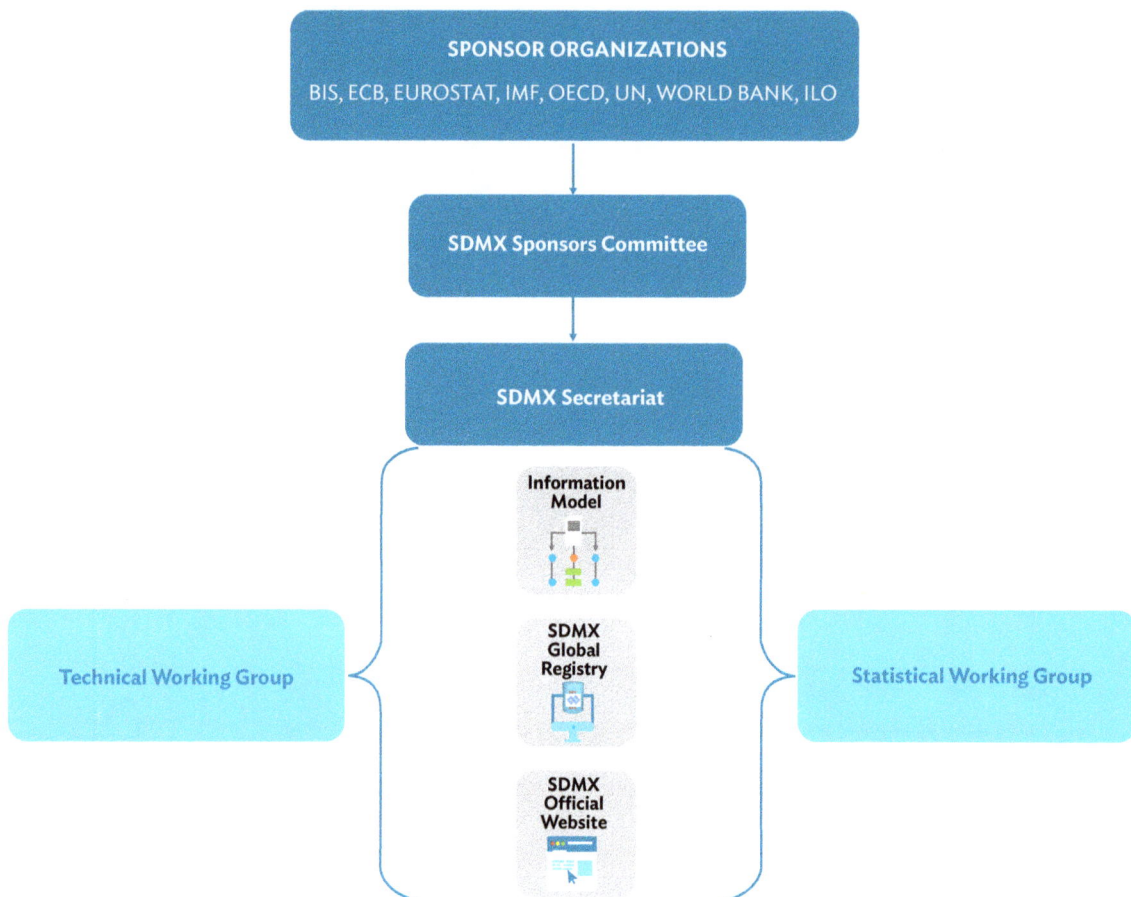

BIS = Bank for International Settlements, ECB = European Central Bank, Eurostat = Statistical Office of the European Union, ILO = International Labour Organization, IMF = International Monetary Fund, OECD = Organisation for Economic Co-operation and Development, SDMX = Statistical Data and Metadata eXchange, UN = United Nations.
Source: Asian Development Bank visualization based on https://sdmx.org/?page_id=2561.

## 3.8 Support from a Global Community of Statistical Practitioners

The adoption of SDMX provides NSOs and other data-producing agencies with access to a global community of individuals and entities actively involved in the development and use of the standard. This opens the way for networking, knowledge exchange, access to wider resources, involvement in standardization efforts, and ongoing professional development. Members of the SDMX community interact frequently to share insights, seek advice, and work together on common challenges related to statistical data management.

In addition, SDMX users have the opportunity to participate in events such as SDMX global conferences, expert meetings organized by the SDMX sponsors, and workshops and webinars hosted by various organizations. These events create a platform for learning and networking on first-hand experiences and best practices in statistical management (SDMX 2020a). Another channel for connecting with the SDMX community of practitioners and experts at any time is the online SDMX User Forum, which was launched in 2022.[5]

---

[5] Access the SDMX User Forum via https://www.yammer.com/unstats/#/home. For more information on SDMX e-learning courses and events, consult the ADB website via https://adb.org or go to any of the following three providers: https://sdmx.org, https://sdmx.io, and https://academy.siscc.org/.

# 4 ISSUES AND CHALLENGES IN NATIONAL STATISTICAL SYSTEMS

Data exchange and dissemination are integral parts of any national statistical system (NSS). Within the NSS, standardized frameworks play a pivotal role in ensuring the reliability, comparability, and coherence of statistical data.

While Statistical Data and Metadata eXchange (SDMX) offers a standardized approach for data exchange and dissemination, its integration into the NSS is sometimes not without challenges. An understanding of these challenges is crucial for statisticians, policymakers, and data practitioners aiming to harness the full potential of SDMX within the complex environment of an NSS (Ward 2015).

## 4.1  Multiple Reporting Organizations

An NSS often involves multiple data-source organizations, each responsible for collecting and reporting specific sets of data. Each organization may have its own data-collection processes and reporting formats (e.g., spreadsheet, text file, digital file), leading to potential inconsistencies and delays in reporting when datasets need to be merged.

The diversity of data-collection methods among agencies often creates resistance to and complexity in standardizing these practices, entrenching difficulties faced in consolidating and harmonizing data from various sources.

Moreover, the need to manage and adapt to multiple reporting formats can be a complex and time-consuming task for NSOs and other data-producing agencies . The need to tailor reports to different formats may require additional resources in terms of time, personnel, and technology.

## 4.2  Different Data Collection Methods

Integrating diverse data sources is critical for ensuring data quality and comparability. However, transforming data collected through different methods into a cohesive and meaningful dataset can be complex.

An NSS may employ different methods of data collection, including censuses, surveys, administrative records, and other sources, and variability in these collection methods can lead to inconsistencies and challenges of integrating data seamlessly.

Managing multiple data-collection methods can also introduce operational complexity for NSOs and other data-producing agencies. Different collection formats, structures, and methodologies may require distinct skill sets, resources, and tools, making full coordination and oversight challenging and expensive.

## 4.3  Lack of Data and Metadata Harmonization

Harmonizing data and metadata from diverse sources involves aligning concepts, classifications, and structures to ensure consistency. Lack of harmonization can result in difficulties in comparing and aggregating data across different statistical domains. It may hinder the interoperability and consistency in the way data are represented and limit the usefulness of statistical information. In addition, the sharing of statistical data becomes more challenging and less efficient.

## 4.4  Manual Data Handling and Validation

Manual data entry, review, and validation processes can be time-consuming and prone to errors. Manual data entry and/or conversion exposes the possibility of inconsistencies in data interpretation and the chance of transcription errors occurring. Similarly, manual data review and validation processes introduce the risk of overlooking inconsistencies and, being slow by nature, they can impact the timeliness of data dissemination.

For instance, when data are sourced from various providers or systems that use different formats (e.g., spreadsheet, text file, digital file), integrating this data manually becomes complex and heightens the chances of errors and inconsistencies being introduced into the dataset.

Identifying and rectifying errors can be challenging, especially in large datasets, leading to the potential misinterpretation of data. Manual data handling and validation therefore have significant potential to compromise the quality and reliability of statistical information.

## 4.5  Unstructured Data

Because unstructured data lack a predefined data model or format, it can be difficult to upload such data into a computerized database.

When unstructured data need to be integrated into an SDMX-enabled database, the process of converting the unstructured data into a structured format becomes challenging. Unstructured data may not adhere to a standardized schema, making it difficult to map the predefined structure expected by the SDMX standard. Moreover, unstructured data may lack clear metadata, requiring additional efforts to define and assign relevant metadata during the mapping process.

# 5 COMPONENTS OF THE STATISTICAL DATA AND METADATA EXCHANGE FRAMEWORK

The Statistical Data and Metadata eXchange (SDMX) standard has three components: a robust information model, content-oriented guidelines, and a detailed technical standard.

The SDMX Information Model describes the key concepts around statistical data, metadata, and data exchange processes. The model can be used to describe any multidimensional dataset, regardless of domain, and is the area of primary focus for statisticians.

The SDMX Content-Oriented Guidelines provide procedures and best practices for creating and managing the content of statistical data and metadata within the SDMX framework. The guidelines offer recommended practices that can be implemented consistently, ensuring interoperability and standardization regardless of the specific focus or subject matter of the statistical data. The SDMX Content-Oriented Guidelines further support the creation of international good practices and shared standards, such as domain-specific data models and cross-domain codelists (sdmx.io n.d.).

The SDMX Technical Standard defines how to create IT tools that fully support the SDMX process throughout the data life cycle. Transforming the SDMX Information Model and the SDMX Content-Oriented Guidelines into tools and databases that respond to the needs of the entire statistical data lifecycle, in accordance with the SDMX Technical Standard, is the focus of IT experts.

## 5.1 Information Model

The SDMX Information Model serves as the central and fundamental component of the SDMX framework. It defines the structure of statistical data and metadata.

More specifically, the model harmonizes the representation and exchange of statistical information across different systems and organizations. It also establishes a common language and structure for expressing statistical concepts to ensure consistency and uniformity in the representation of data. The model identifies objects within the statistical domain and defines their relationships. This includes essential elements such as core concepts, their roles, and the codelists that enable a clear understanding of the interconnections between various statistical agencies.

Moreover, the model provides a standardized approach to organizing and accessing statistical data. This contributes to interoperability, allows for centralized management, and simplifies the process involved in data exchange across different platforms and systems (Eurostat SDMX Infospace).

## 5.1.1 Key Elements of the Information Model

The SDMX Information Model comprises, among others, the following key elements:

(i)    descriptor concepts (i.e., concepts associated with the statistical data) as well as the nature of these concepts (dimension, attribute, or measure);

(ii)   the packaging structure (i.e., observation level, series level, dataset level);

(iii)  the keys (grouping the various dimensions for a particular set of data); and

(iv)   the codelist (defining the possible values for a dimension).

As shown in Figure 7, all the information for the key elements is contained in a specific data structure definition (DSD) or "key family".

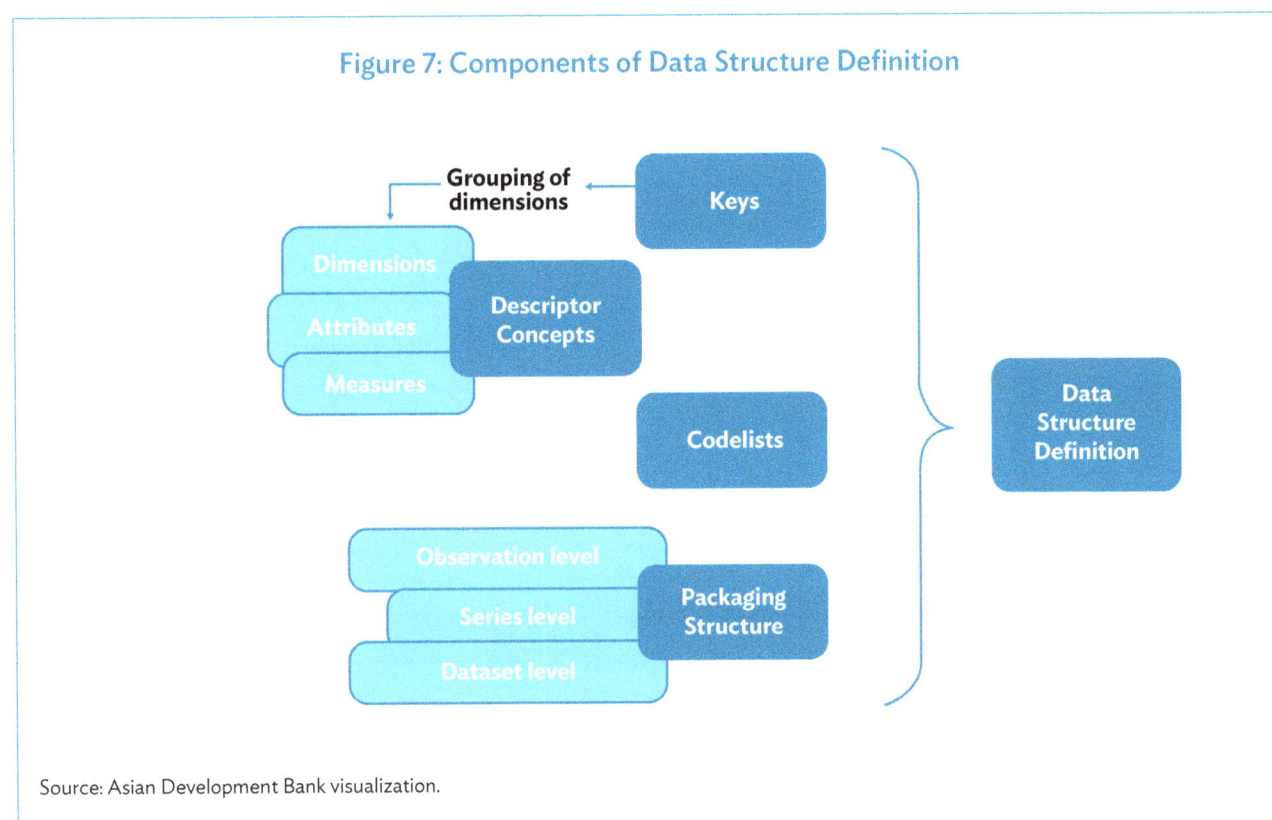

Figure 7: Components of Data Structure Definition

Source: Asian Development Bank visualization.

The DSD specifies a set of concepts—also referred to as "structural metadata"—to describe and identify a set of data. A statistical concept in SDMX refers to a unit of thought created by a unique combination of characteristics (SDMX 2020c). It provides essential information about the data, such as the location or economy represented (e.g., reference area), the specific time period to which observation refers (e.g., time period), and the statistical aspect to which the data pertain (e.g., indicator).

The concept, codelist, and packaging structure are essential components of any DSD. The keys shown in Figure 7 refer to the combination of dimension values that uniquely identifies an observation or series within a dataset (SDMX 2020c). In the case of a time series, the keys include all dimensions except time period.

Every dataset is defined using three concepts from the SDMX Information Model. The first concept is **dimension**, which is used for defining the data: dimensions are always categorical, meaning most of the dimensions have a codelist except for time period which follows a structured format such as "YYYY" for yearly data, "YYYY-QQ" for quarterly data, "YYYY-MM" for monthly, and more.[6] The second concept is **attribute**, which is used only for describing additional aspects of the data: common examples of attributes are footnotes and other descriptive text. The final concept is **measure**, which is numerical and represents an actual value.

In other words, a DSD is a container for metadata that describe the structure of related datasets in terms of their dimensions, attributes, and measures (sdmx.io n.d.).

Additional explanatory information is called "reference metadata", with this information describing the content, methodology, and quality of the data. In SDMX terminology, the data and metadata structure definitions are made available in the SDMX Global Registry (Tissot 2018).

As an example, Table 1 provides summary of the Economic and Financial Statistics Data Structure Definition developed and maintained by the  International Monetary Fund.

#### Table 1: Summary of the Economic and Financial Statistics Data Structure Definition

| CONCEPT | | | CODELIST | PACKAGING STRUCTURE |
|---|---|---|---|---|
| ID | NAME | ROLE | | |
| DATA_DOMAIN | Data domain | Dimension | CL_DATADOMAIN | |
| REF_AREA | Reference country or area | Dimension | CL_REF_AREA | |
| INDICATOR | Economic indicator | Dimension | CL_INDICATOR | |
| COUNTERPART_AREA | Counterpart area | Dimension | CL_REF_AREA | |
| FREQ | Frequency | Dimension | CL_FREQ | |
| TIME_PERIOD | Time period | Dimension | | |
| BASE_PER | Base period | Attribute | | Series level |
| UNIT_MULT | Unit multiplier | Attribute | CL_UNIT_MULT | Series level |
| TIME_FORMAT | Time format | Attribute | CL_TIME_FORMAT | Series level |
| COMMENT | Comment | Attribute | | Dataset level |
| OBS_STATUS | Observation status | Attribute | CL_OBS_STATUS | Observation level |
| OBS_VALUE | Observation value | Measure | | |

CL = codelist, FREQ = frequency, ID = unique identification of the concept, MULT = multiplier, OBS = observation, PER = period, REF = reference.
Note: The Economic and Financial Statistics Data Structure Definition was developed and is maintained by the International Monetary Fund.
Source: Asian Development Bank construction based on the Economic and Financial Statistics Data Structure Definition taken from https://sdmxcentral.imf.org/data/datastructure.html.

---

6    For example, data for the year 2024 is formatted as "YYYY": 2024; data for the first quarter of 2024 is formatted as "YYYY-QQ": 2024-Q1; data for May 2024 is formatted as "YYYY-MM": 2024-05.

## 5.1.2  The Importance of Structural Data Modeling

Statistics are generally concerned with particular details about objects or events. A few examples are persons, households, geographic areas, loans, jobs, marriages, and births.  Objects and events have a variety of characteristics that help define or describe them. A few characteristics of a person, for example, are age, weight, height, eye color, country of birth, education status, employment status, nationality, mother tongue, and languages spoken.

Data modeling is a general methodology for deciding on and defining all the objects and events to be used in a dataset, determining the characteristics needed to accurately define and describe the statistical data, and specifying the relationships between the data and the processes that act upon them.

Structural data modeling, more particularly, is a systematic approach to conceptualizing and organizing information within a specific context. It involves the clear and unambiguous identification and description of concepts, the selection of key properties and specification of their attributes, the definition of the relationships between concepts, and the formal codification of these concepts (Figure 8). This process is fundamental in various fields—including database design, systems analysis, and information management— as it provides a clear and structured representation of the data within a given domain (UNDESA 2019).

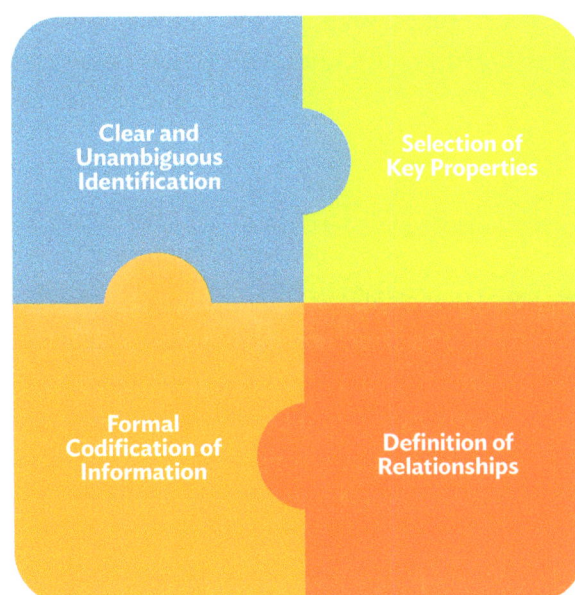

Figure 8 : The Four Quadrants of Structural Data Modeling

Clear and Unambiguous Identification

Selection of Key Properties

Formal Codification of Information

Definition of Relationships

Source: Asian Development Bank visualization.

**Clear and unambiguous identification** involves carefully identifying and delineating the concepts to describe the data. This involves a precise definition of what these concepts represent in the context of the data. For example, age disaggregation is not relevant in the definition for international merchandise trade statistics but is pertinent for social indicators.

**Selecting key properties** means accurately identifying the essential properties that need to be captured to describe the concepts meaningfully. These properties provide crucial information about the characteristics of the identified concepts. Examples of such properties include a unique identifier, a name, and a description.

For instance, within the concept of "frequency", the identifier is : "FREQ", the name is "frequency of observation", and the description is "time interval at which observations occur over a given time period". Additionally, name and description can be expressed in multiple languages, with English as the default.

**The definition of relationships** between concepts requires understanding and formalizing connections within the data. For statistics to be of high quality and comparable, the model used to define and describe the data needs to be standardized, with common concepts specified using common terms. In a world with many languages and alphabets, the most effective approach to this challenging task is to use codes to represent the concepts. A code is a language-independent set of letters, numbers, or symbols that represent a concept, the meaning of which is described in a natural language. For example, under the concept of sex or gender, the code "F" can represent female (English), femme (French), or femenina (Spanish).

**Formal codification of information** involves creating a structured representation of all concept codes by, for example, creating codelists for all categorical variables, e.g., sex, age, eye color, or employment status, and grouping relevant concepts together into data structures. This enables ease of management and dataflows to support processes such as sharing data with other organizations, publishing in a data portal, using data within internal processes, and collecting data from providers.

## 5.1.3  Structural Data Modeling in the Context of the Standard

Structural data modeling is an essential part of SDMX. It ensures a standardized and consistent representation of statistical information. The structural data model of SDMX serves as the foundation for organizing, describing, and exchanging data and metadata across diverse systems and organizations. It can vary across different statistical domains as some data concepts may be essential in one domain but not relevant in another. For instance, while sex may not be relevant in a consumer price index, it holds significance for many Sustainable Development Goal indicators.

By identifying and defining statistical concepts, specifying data structures, and allowing interoperability, structural data modeling in SDMX promotes seamless data exchange, consistency, and the efficient implementation of the SDMX standard. This form of data modeling plays a crucial role in quality assurance, metadata description, and the transformation of datasets, making it an integral aspect of SDMX's goal to provide a universal framework for statistical data exchange.

## 5.1.4 Message Formats under the Standard

Message formats provide standardized ways to structure and convey statistical information between systems. The choice of the format depends on factors such as data complexity, interoperability requirements, and the specific needs of data producers and consumers. SDMX supports multiple message formats—including Extensible Markup Language (XML), JavaScript Object Notation (JSON), and comma-separated values (CSV)—to cater to diverse user needs and preferences.

### Extensible Markup Language

The default message format of SDMX is in Extensible Markup Language, known as SDMX-XML or SDMX-ML. A sample of the message format is shown in Figure 9. XML is a robust and widely adopted format, offering a hierarchical and extensible structure. It provides a standardized way to represent complex relationships within statistical data and metadata, ensuring consistency and ease of interpretation by both humans and machines. XML's extensibility allows for the inclusion of additional information or customizations, assisting adaptation to specific needs while maintaining compatibility.

**Figure 9: Sample Message Using Extensible Markup Language**

This XML file does not appear to have any style information associated with it. The document tree is shown below.

```
▼<message:StructureSpecificData xmlns:ss="http://www.sdmx.org/resources/sdmxml/schemas/v2_1/data/structurespecific"
  xmlns:message="http://www.sdmx.org/resources/sdmxml/schemas/v2_1/message"
  xmlns:generic="http://www.sdmx.org/resources/sdmxml/schemas/v2_1/data/generic"
  xmlns:footer="http://www.sdmx.org/resources/sdmxml/schemas/v2_1/message/footer"
  xmlns:common="http://www.sdmx.org/resources/sdmxml/schemas/v2_1/common" xmlns:xsi="http://www.w3.org/2001/XMLSchema-instance">
    <script/>
  ▼<message:Header>
      <message:ID>KI</message:ID>
      <message:Test>false</message:Test>
      <message:Prepared>2024-06-13 12:23:47</message:Prepared>
    ▼<message:Sender id="ADB">
        <common:Name xml:lang="en">Asian Development Bank</common:Name>
      ▼<message:Contact>
          <common:Name xml:lang="en">SDMX API Admin</common:Name>
          <message:Email>jdelacruz3.consultant@adb.org</message:Email>
        </message:Contact>
      </message:Sender>
      <message:DataSetAction>Information</message:DataSetAction>
      <message:DataSetID>IAEG-SDGs</message:DataSetID>
      <message:Extracted text="2024-06-13 12:23:47"/>
      <message:Records totalRecords="11" recordsPerPage="10000" totalPages="1" currentPage="1"/>
    </message:Header>
  ▼<message:DataSet action="Append" structureRef="DF_SDG_GLH">
    ▼<Series REF_AREA="THA" INDICATOR="SI_POV_EMP1_15T" FREQ="A" UNIT="ratio" UNIT_MULT="Unit" REF_YEAR="" BASE_YEAR=""
      METHODOLOGY="">
        <Obs TIME_PERIOD="2005" OBS_VALUE="0.67"/>
        <Obs TIME_PERIOD="2006" OBS_VALUE="0.72"/>
        <Obs TIME_PERIOD="2007" OBS_VALUE="0.4"/>
        <Obs TIME_PERIOD="2008" OBS_VALUE="0.16"/>
        <Obs TIME_PERIOD="2009" OBS_VALUE="0.2"/>
        <Obs TIME_PERIOD="2010" OBS_VALUE="0.11"/>
        <Obs TIME_PERIOD="2011" OBS_VALUE="0.05"/>
        <Obs TIME_PERIOD="2012" OBS_VALUE="0.07"/>
        <Obs TIME_PERIOD="2013" OBS_VALUE="0.05"/>
        <Obs TIME_PERIOD="2014" OBS_VALUE="0.01"/>
        <Obs TIME_PERIOD="2015" OBS_VALUE="0.01"/>
      </Series>
    </message:DataSet>
  </message:StructureSpecificData>
```

Source: Asian Development Bank screenshot taken from internal systems..

## JavaScript Object Notation

JSON is a lightweight data interchange format that is easy for humans to read and write, and easy for machines to parse and generate. SDMX-JSON is a specific message format designed for the exchange of statistical data and metadata in accordance with the SDMX standard. A sample of the format is shown in Figure 10. SDMX-JSON aligns with modern web development practices, making it suitable for web services (e.g., application programming interfaces) and applications that operate in a JSON-centric environment.

### Figure 10: Sample Message Using JavaScript Object Notation

```
"structure": {
  "name": "IAEG-SDGs",
  "description": "Inter-agency and Expert Group on SDG Indicators",
  "dimensions": {
    "series": [
      {
        "id": "REF_AREA",
        "name": "Economy",
        "values": [
          {
            "id": "THA",
            "name": null
          }
        ]
      },
      {
        "id": "SUBJECT",
        "name": "Subject",
        "values": {
          "id": "SI_POV_EMP1_15T",
          "name": "PPP - Intl Poverty Line- 15+ Total",
          "description": null
        }
      },
      {
        "id": "FREQ",
        "name": "Frequency",
        "values": {
          "id": "A",
          "name": "Annual"
        }
      }
    }
  }
```

Source: Asian Development Bank screenshot taken from internal systems.

## Comma-Separated Values

CSV is a plain-text format in which values are separated by commas and each line represents a record. The tabular structure of CSV aligns well with the representation of tabular statistical data, making it suitable for datasets organized in rows and columns. SDMX-CSV is a particular message format used for the exchange of statistical data and metadata following the SDMX standard. A sample of this format is shown in Figure 11.

Figure 11: Sample Message Using Comma-Separated Values

| | A | B | C | D | E | F | G | H |
|---|---|---|---|---|---|---|---|---|
| | Dataflow | SERIESKEY | FREQ | REF_AREA | TIME_PERIOD | OBS_VALUE | UNIT | UNIT_MULT |
| | IAEG-SDGs:DF_SDG_GLH | SI_POV_EMP1_15T | A:Annual | THA:Thailand | 2005 | 0.67 | PT:Percent | 0:Unit |
| | IAEG-SDGs:DF_SDG_GLH | SI_POV_EMP1_15T | A:Annual | THA:Thailand | 2006 | 0.72 | PT:Percent | 0:Unit |
| | IAEG-SDGs:DF_SDG_GLH | SI_POV_EMP1_15T | A:Annual | THA:Thailand | 2007 | 0.4 | PT:Percent | 0:Unit |
| | IAEG-SDGs:DF_SDG_GLH | SI_POV_EMP1_15T | A:Annual | THA:Thailand | 2008 | 0.16 | PT:Percent | 0:Unit |
| | IAEG-SDGs:DF_SDG_GLH | SI_POV_EMP1_15T | A:Annual | THA:Thailand | 2009 | 0.2 | PT:Percent | 0:Unit |
| | IAEG-SDGs:DF_SDG_GLH | SI_POV_EMP1_15T | A:Annual | THA:Thailand | 2010 | 0.11 | PT:Percent | 0:Unit |
| | IAEG-SDGs:DF_SDG_GLH | SI_POV_EMP1_15T | A:Annual | THA:Thailand | 2011 | 0.05 | PT:Percent | 0:Unit |
| | IAEG-SDGs:DF_SDG_GLH | SI_POV_EMP1_15T | A:Annual | THA:Thailand | 2012 | 0.07 | PT:Percent | 0:Unit |
| | IAEG-SDGs:DF_SDG_GLH | SI_POV_EMP1_15T | A:Annual | THA:Thailand | 2013 | 0.05 | PT:Percent | 0:Unit |
| | IAEG-SDGs:DF_SDG_GLH | SI_POV_EMP1_15T | A:Annual | THA:Thailand | 2014 | 0.01 | PT:Percent | 0:Unit |
| | IAEG-SDGs:DF_SDG_GLH | SI_POV_EMP1_15T | A:Annual | THA:Thailand | 2015 | 0.01 | PT:Percent | 0:Unit |

Source: Asian Development Bank screenshot taken from internal systems.

## 5.2 Content-Oriented Guidelines

The SDMX Content-Oriented Guidelines (SDMX-COGs) refer to recommended practices for the creation of interoperable statistical data and metadata sets (SDMX 2016).

A primary focus of the SDMX-COGs is harmonizing specific concepts and terminologies that are common to a large number of statistical domains. By aligning concepts and terminology, the SDMX-COGs create a shared language and understanding. The harmonization promoted by the SDMX-COGs contributes to a more efficient exchange of comparable data and metadata. This is particularly valuable where datasets need to be compared or combined across different statistical domains: common terminology and concepts enhance the consistency and reliability of such comparisons.

The SDMX-COGs are developed using implementation experiences within the SDMX community. They leverages lessons learned from practical implementations, ensuring that the guidelines are grounded in real-world scenarios and are reflective of best practices globally.

### 5.2.1 Cross-Domain Concepts

A statistical concept in SDMX describes all relevant characteristics of the data, each assigned to represent as either a dimension, attribute, or measure. The term "cross-domain" is used to indicate that a concept can be applied in more than one statistical domain in a materially similar form. In other words, these concepts exhibit a substantial degree of similarity when applied in different domains.

Figure 12 illustrates two distinct statistical domains. In Domain 1, statistical concepts like customs procedure, trade flow, trade system, reference area, time period, and observation value are used. Meanwhile, Domain 2 uses concepts such as accounting entry and stocks, alongside an activity classification. Notably, both domains share certain concepts that are presented in a materially similar form. These include reference area, time period, and observation value. These shared concepts serve as prime examples of cross-domain concepts.

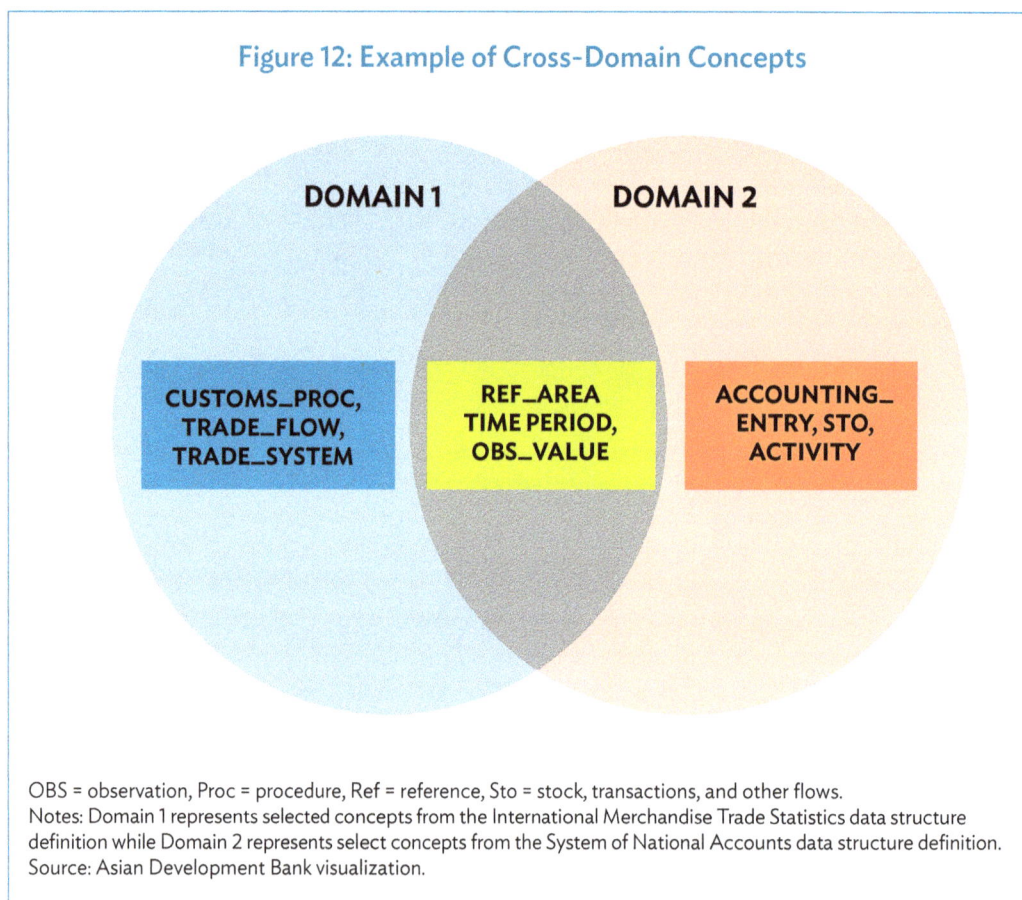

## Figure 12: Example of Cross-Domain Concepts

**DOMAIN 1**          **DOMAIN 2**

CUSTOMS_PROC, TRADE_FLOW, TRADE_SYSTEM

REF_AREA TIME PERIOD, OBS_VALUE

ACCOUNTING_ ENTRY, STO, ACTIVITY

OBS = observation, Proc = procedure, Ref = reference, Sto = stock, transactions, and other flows.
Notes: Domain 1 represents selected concepts from the International Merchandise Trade Statistics data structure definition while Domain 2 represents select concepts from the System of National Accounts data structure definition.
Source: Asian Development Bank visualization.

Cross-domain concepts in the SDMX framework refer to concepts that are relevant to many, if not all, statistical domains (SDMX 2016). These are fundamental elements describing statistical data and are intended for use across different domains wherever possible .

The SDMX cross-domain concepts are actively developed and published by the SDMX Statistical Working Group. This group is responsible for shaping standardization efforts and ensuring that cross-domain concepts are well-defined, widely applicable, and adhere to best practices.

## 5.2.2  Codes and Codelists

Codes that are language-independent and used globally are crucial for ensuring standardized communication and data interchange across diverse systems and languages. For instance, in the aviation and travel industries, three-letter codes provide a standardized and globally recognized way of identifying airports (e.g., MNL for Manila airport, SIN for Singapore airport). These codes are used on flight information displays at airports and on airline websites. Travelers can quickly locate their departure and arrival gates, check flight status, and navigate airports more efficiently using these codes.

Each code within a codelist should be well-described. This involves providing clear and comprehensive information about the meaning, context, and usage of each code. Well-documented codes contribute to the understanding and correct interpretation of statistical data.

Codelists are established to organize interrelated concept codes into a meaningful, systematic, and standardized format. The aim is to establish a structured and standardized format for organizing terms that represent specific concepts in statistical data and metadata. A codelist functions as a collection of codes maintained as a unit.

As part of the SDMX-COGs, codelists refer to predefined sets of terms from which certain statistically coded concepts derive their values (SDMX 2018). To provide SDMX codelists with a distinct visual identity, it is recommended to prefix their identifiers with **CL_**. The codelist comprises three essential elements: an identification number, a version number, and a reference to a maintenance agency.

Table 2 provides a sample of cross-domain SDMX codelists.

### Table 2: Sample of Cross-Domain Codelists

| Id | Name | Description | Agency Id | Version |
|---|---|---|---|---|
| CL_FREQ | Frequency | This codelist provides a set of values indicating the "frequency" of the data (e.g. weekly, monthly, quarterly). The concept "frequency" may refer to various stages in the production process, e.g. data collection or data dissemination. For example, a time series could be disseminated at annual frequency but the underlying data are compiled monthly. The codelist is applicable for all different uses of "frequency". | SDMX | 2.1 |
| CL_UNIT_MULT | Unit multiplier | This codelist provides code values for indicating the magnitude in the units of measurement. More information about this codelist and SDMX codelists in general (e.g. list of generic codes for expressing general concepts like "Total", "Unknown", etc.; syntaxes for the creation of further codes; general guidelines for the creation of SDMX codelists) can be found at this address: https://sdmx.org/?page_id=4345. | SDMX | 1.1 |
| CL_SERIES | SDG series codelist | | IAEG-SDGs | 1.6 |
| CL_EDUCATION_LEV | SDG educational level codelist | | IAEG-SDGs | 1.0 |
| CL_CURRENCY | Currency of issuance or invoicing codelist | | IMF | 1.6 |

CL = codelist, Freq = frequency, IAEG-SDGs = Inter-agency and Expert Group on Sustainable Development Goal Indicators, IMF = International Monetary Fund, Lev = level, Mult = multiplier, SDMX = Statistical Data and Metadata eXchange.
Note: Name and Description can be expressed in multiple languages.
Source: Asian Development Bank screenshot taken from SDMX foundation course material.

### 5.2.3 Statistical Subject-Matter Domain

The statistical subject-matter domain is a fundamental concept within the SDMX-COGs, emphasizing the common characteristics and standardized approaches associated with specific statistical activities (SDMX 2016). The SDMX framework provides structured classifications for organizing data based on these subject-matter domains such as national accounts, balance of payments, international merchandise trade, consumer price indices, and Sustainable Development Goals (SDGs), among others.

Classifications under statistical subject-matter domains provide a high-level scheme for organizing statistical data and metadata in various types of applications. These classifications are based on the United Nations Economic Commission for Europe's Classification of International Statistical Activities.

The classifications are often expressed as data structure definitions (DSDs). The SDMX sponsor organizations have defined global DSDs for various statistical domains, including consumer price indices, national accounts, balance of payments, international merchandise trade statistics, the SDGs, the System of Environmental-Economic Accounting, government finance statistics, and foreign direct investment. Individual economies have the flexibility to customize existing DSDs to align with their national data needs, with the scope to expand the codes to accommodate their particular statistical objectives. The established global DSDs can be found in the SDMX Global Registry.[7] When reporting SDG data, national statistics offices use the SDG global DSD to report to international organizations.

### 5.2.4 Glossary

The SDMX Glossary contains concepts and related definitions used in structural and reference metadata across international organizations and national data-producing agencies (SDMX 2020c). It focuses on terms essential for constructing and understanding metadata systems and facilitating SDMX data exchange arrangements.

Rather than imposing specific concepts and codelists for SDMX structures, the glossary recommends a common terminology. This recommendation aims to foster communication and understanding by suggesting a shared vocabulary while allowing flexibility in the use of concepts and codes. Figure 13 provides an overview of the cross-domain concept incorporated in the SDMX Glossary.

---

[7] Access the global DSDs in the SDMX Global Registry via https://registry.sdmx.org/data/datastructure.html.

### Figure 13: Cross-Domain Concept in the Glossary

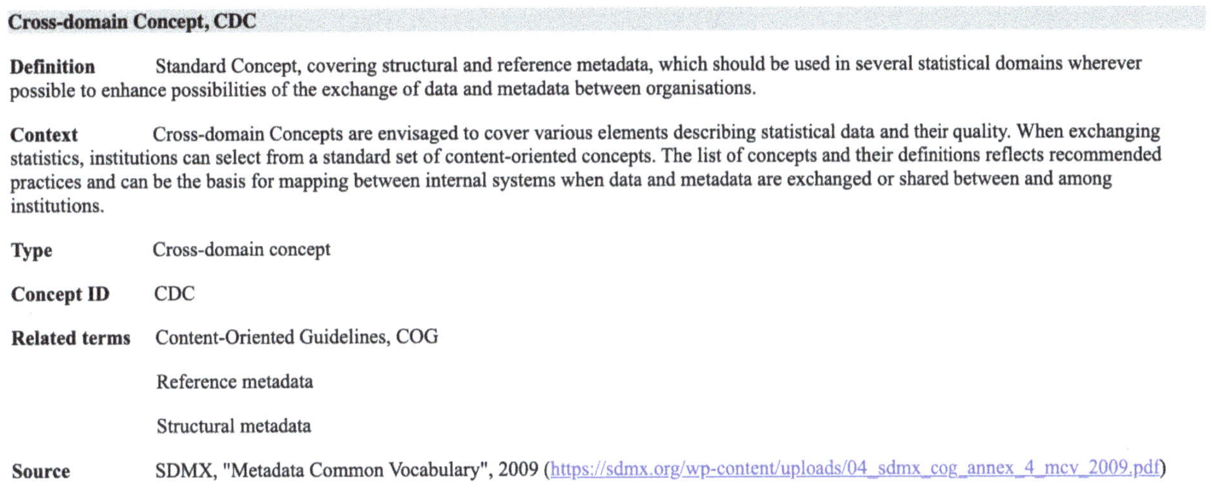

**Cross-domain Concept, CDC**

**Definition**        Standard Concept, covering structural and reference metadata, which should be used in several statistical domains wherever possible to enhance possibilities of the exchange of data and metadata between organisations.

**Context**        Cross-domain Concepts are envisaged to cover various elements describing statistical data and their quality. When exchanging statistics, institutions can select from a standard set of content-oriented concepts. The list of concepts and their definitions reflects recommended practices and can be the basis for mapping between internal systems when data and metadata are exchanged or shared between and among institutions.

**Type**        Cross-domain concept

**Concept ID**        CDC

**Related terms**        Content-Oriented Guidelines, COG

Reference metadata

Structural metadata

**Source**        SDMX, "Metadata Common Vocabulary", 2009 (https://sdmx.org/wp-content/uploads/04_sdmx_cog_annex_4_mcv_2009.pdf)

Source: Asian Development Bank screenshot taken from https://sdmx.org/wp-content/uploads/SDMX_Glossary_Version_2_1_December_2020.htm#_Toc59116750.

## 5.3  Technical Standard

For some time, statistics offices have used information technology (IT) to improve the efficiency of statistical processes. However, such innovations have remained confined only to specialized areas, creating islands where IT tools are less likely to interoperate, share data, and work together as a cohesive system. This fragmented environment of individual and disparate IT systems created what many researchers called "silos" within statistics offices.

A data flow analysis offers active measures that statistics offices can take to break the silos by harnessing data in digital format to realize the gains of using advanced technologies (Paris21 2021).

Data flow analyses can also pave the way for statistics producers to follow the SDMX Technical Standard, which defines the configuration of IT tools to support the SDMX process through the entire data life cycle. The technical standard transforms the SDMX Information Model and the SDMX Content-Oriented Guidelines into practical systems and databases, including web services application programming interface (API) specification, transmission format specifications, and the SDMX registry specification.

The web services API specification provides a standardized interface for interacting with software systems implementing the SDMX standard. An organization's SDMX-API web service typically would offer programmatic access to data and metadata published on the organization's data portal. The related open API documentation describes the supported functionality in an interactive way. Data retrieval and discovery are supported in a variety of formats (e.g., JSON, XML, CSV). Data and structural metadata such as DSDs, codelists, and concept schemes are available via this API service. The latest version of the standard,

SDMX 3.0, supports transmission formats of data and structural metadata in XML and JSON formats, and of data only in CSV format. The XML and JSON components also support reference metadata. The SDMX registry is a controlled repository for structural metadata and processes, which organizations can consult for information on how to structure, process, validate, and interpret statistical data.

# 6 STEPS TO IMPLEMENTING STATISTICAL DATA AND METADATA EXCHANGE

The adoption of the Statistical Data and Metadata eXchange (SDMX) standard should be thought of as a change project or, depending on the quantity and breadth of statistical data and activities across the organization, many change projects. These projects will typically span multiple years and impact statisticians, economists, communications, and information technology (IT) staff within an organization.

## 6.1 Identifying Organizational Priorities and Applications of the Standard

Statistical organizations typically use a variety of approaches to decide where to invest scarce financial and human resources. Needs may be identified as a result of audits on data quality and/or governance, an organizational assessment of risks and opportunities, or a maturity assessment using a data quality framework. In other instances, there may be a desire to implement a significant new capability, such as a data portal, or the need to accommodate an international initiative, such as the Sustainable Development Goals. The first step in adopting the SDMX standard is therefore to undertake a rigorous assessment and determine alignment of SDMX capabilities with organizational goals and priorities. This assessment should then be followed by the development of a prioritized road map of projects to guide the SDMX adoption efforts. Paris21, which was established to advance statistical capacity in low- and middle-income economies, has defined a data flow assessment framework to assist organizations with this process. The Paris21 guidelines help statistical organizations assess and document data flows, specifying how data is collected, processed, and disseminated. The goal during this phase is to become familiar with the fundamental concepts of SDMX and to identify and prioritize the primary applications of SDMX within the organization, such as data reporting to international organizations using global data structure definitions (DSDs). In other cases, the organization may already have decided on their priority projects, such as implementing a data portal, and the change effort is focused on specific uses of SDMX.

## 6.2 Developing a Strategy for Organizational Upskilling

As the depth and scope of data being modeled increase, so must the data governance practices that ensure the quality, coherence, and consistency of the model and the modeled artefacts.

The second step in SDMX adoption is therefore to establish a resourcing strategy for upskilling staff to become SDMX literate. These staff members must have the requisite knowledge and skills to model data in accordance with the SDMX standard and its recommended modeling practices.

A governance framework for SDMX implementation typically would establish roles, responsibilities, and data quality assurance procedures. The goals during this phase are to broaden and deepen the knowledge of SDMX among key staff, especially the statisticians who are responsible for data modeling; and to create the necessary SDMX information model artefacts for the project, such as DSDs, including the concept scheme, codelists, and roles.[8]

## 6.3 Understanding the Technology Needs Under Implementation

In parallel with data modeling, a stream of work should be undertaken to assess the IT needs to deliver on the initial projects as well as the architectural implications for delivering on the full range of projects that have been identified.

It is important during this IT and architecture phase to fully understand the organization's vision for SDMX adoption. While changing tools later on in the adoption process is feasible, selecting the tools that are the best fit from the very beginning generally results in the most efficient use of resources. Changing tools midway through an SDMX adoption creates data quality and project delivery risks, but this may be the best approach in certain circumstances. These deliberations should involve discussions with other organizations who have successfully implemented SDMX as well as accessing other educational resources.[9]

The goal for this stream of work is to upskill the core team to be adept SDMX practitioners in the modeling of statistical data and to have produced the necessary SDMX artefacts for the first SDMX application within the organization.

## 6.4 Other Factors for Consideration

As with any change initiative, a communication strategy will be required to engage internal and external stakeholders affected by the SDMX adoption project (or projects) and a robust testing and transition processes should be undertaken.

It is desirable, and in most cases quite feasible, to realize concrete benefits in data quality and operational efficiency throughout the phases of SDMX implementation.

---

[8] There are a number of resources available to assist organizations in this regard, such as the ADB e-learning course on SDMX Foundation (https://elearn.adb.org/course/view.php?id=486) and SDMX Tools (https://elearn.adb.org/course/view.php?id=520), the .Stat academy (https://academy.siscc.org/), and sdmx.io (https://sdmx.io/resources/elearning). For a current list of all SDMX courses, consult the learning catalogue on the sdmx.org website.

[9] These resources include ADB's e-learning course on SDMX tools, the sdmx.io site managed by the Bank for International Settlements, and the sdmx.org tools site.

Once the organization has achieved its initial goals in adopting SDMX, more advanced implementations can be considered. This could involve extending the adoption of SDMX to other sectors (domains) within the organization or extending the adoption of SDMX to other statistical processes such as data collection or production. Advanced implementations could enable the organization to automate data exchange processes using SDMX-compliant application programming interfaces, or to further improve data quality by implementing advanced data and metadata validations and transformations for via extensions such as the Validation and Transformation Language.

# 7 IMPORTANT APPLICATIONS OF STATISTICAL DATA AND METADATA EXCHANGE

Two primary applications of Statistical Data and Metadata eXchange (SDMX) for individual economies are the National Summary Data Page (NSDP) and the more general statistical data portal.

## 7.1 National Summary Data Page

The NSDP functions as a data dissemination platform for economies actively engaged in the International Monetary Fund's Special Data Dissemination Standard (SDDS), SDDS Plus, or the Enhanced General Data Dissemination System (e-GDDS).

As the name suggests, the NSDP is a web page that provides a summary of key macroeconomic and financial data and social indicators for the economy involved in SDDS, SDDS Plus, or e-GDDS (IMF n.d.).[10] The NSDP aims to enhance data dissemination and improve the accessibility and visibility of the data for end users. It provides access to a diverse range of online data and metadata across all available categories for a given economy. Notably, even where these categories are compiled by multiple statistical agencies (e.g., national statistics offices, central banks, etc.), the NSDP offers a simple way of disseminating data in SDMX format for various statistical domains (IMF n.d.) where economies are engaged in SDDS Plus or e-GDDS.

The NSDP is hosted and maintained by the national statistics office or other relevant government agency. It serves as a great resource for policymakers, researchers, and the general public to access important statistical information.

Figure 14 shows an example of Bhutan's NSDP in SDMX format.[11]

---

[10] As of 14 June 2024, 36 ADB members in Asia and the Pacific were participating in SDDS, SDDS Plus, and e-GDDS.

[11] Access the current NSDP of Bhutan via https://www.nsb.gov.bt/national-summary-data-page-nsdp-bhutan/.

Figure 14: National Summary Data Page for Bhutan

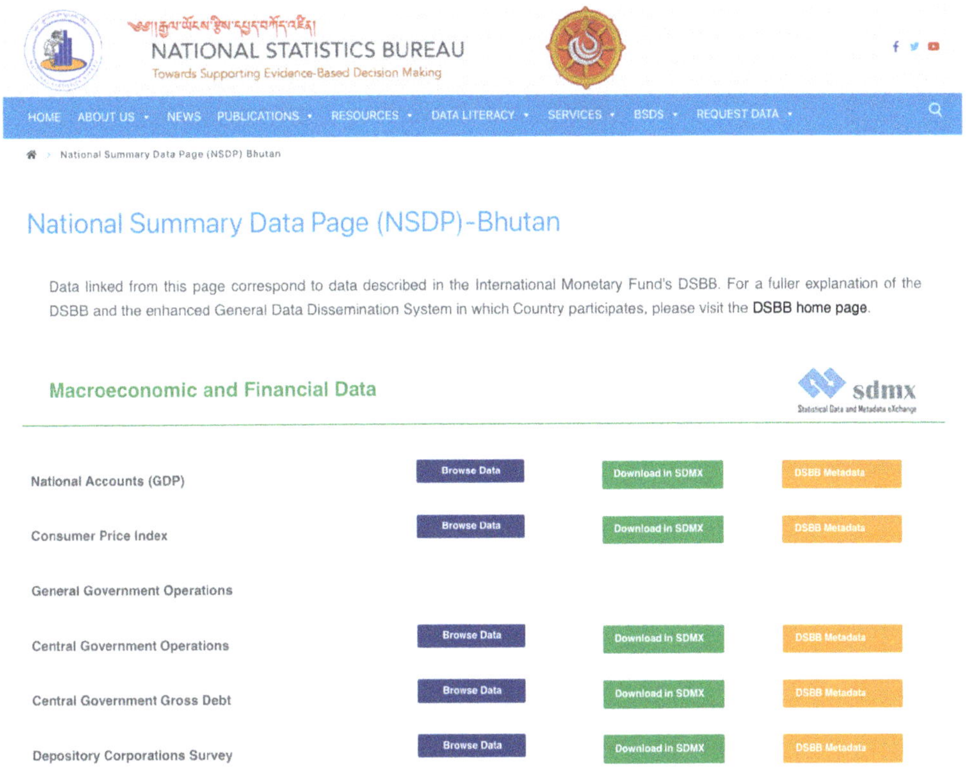

Source: Asian Development Bank screenshot taken from https://www.nsb.gov.bt/national-summary-data-page-nsdp-bhutan/.

## 7.2  Data Portal

With an increasing need for efficient data sharing, the online data portal serves as one of the most advanced models for disseminating information. Such platforms are designed to simplify the exploration, retrieval, and visualization of statistical data.

The Asian Development Bank (ADB), through the Data Division of the Economic Research and Development Impact Department, supports National Statistics Offices in the region in adopting the of the SDMX standard. A prime example of a quality online data portal using SDMX is the Statistics Sharing Hub developed and maintained by the National Statistical Office of Thailand (TNSO).

The statistical system in Thailand functions in a decentralized manner, whereby various government agencies independently undertake statistical activities aligned with their respective missions, which lead to dispersion of diverse statistical data in both the public and private sectors. Moreover, the different agencies involved employ diverse formats and definitions for data storage. The challenge therefore lies in integrating these datasets for effective public administration or crisis response.

The solution to this challenge required the establishment of standardized methods for the exchange of statistical data, including the development of a statistical framework that defined common data formats and terminology. Moreover, this framework needed to allow automation of data exchange, thereby streamlining the process of data integration and enhancing the efficiency of data utilization.

The TNSO took the initiative to establish a robust platform capable of integrating diverse indicators into a centralized dissemination hub featuring advanced functionalities and SDMX compliance. These ambitions led to the creation of the TNSO's Statistics Sharing Hub, powered by the .Stat Suite—a potent SDMX tool renowned for its effectiveness in disseminating statistical data and metadata.

Moreover, the TNSO also used a combination of tools to implement the SDMX standard. For structural metadata repository, the office used the Fusion Metadata Registry to store, publish, and maintain their SDMX artefacts such as the DSDs, dataflows, concept schemes, and codelists. For SDMX data preparation, the TNSO used Excel2CSV to convert to SDMX-CSV format and uploaded to .Stat Suite for data dissemination (TNSO 2023).

Box 1 provides more information about the TNSO's implementation of SDMX, while Box 2 and Box 3 showcase benefits gained from SDMX implementations in Maldives and across the Pacific.

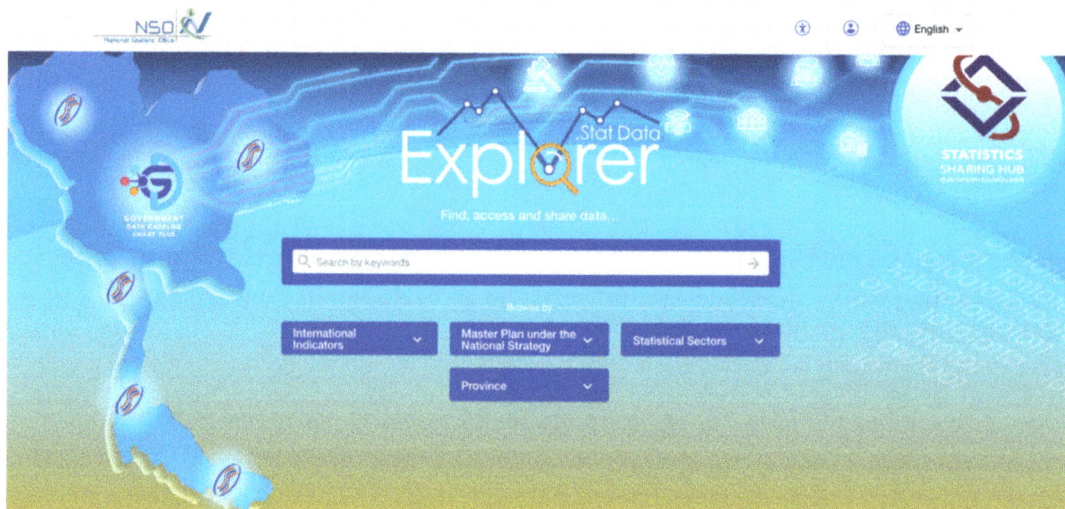

**Box 1: Streamlining Statistics Sharing in Thailand**

Source: Screenshot of the National Statistical Office of Thailand's Statistical Sharing Hub taken from https://stathub.nso.go.th/?lc=en&pg=0.

*continued on next page*

Box 1: *continued*

## Background

The statistical system in Thailand operates in a decentralized manner, with government agencies conducting various independent statistical activities aligned with their missions. This results in diverse statistical data scattered across the public and private sectors. While there is an abundance of data, integrating diverse datasets for public administration or crisis response is challenging due to variations in data storage formats and definitions among different agencies.

To address this issue, the National Statistical Office of Thailand (TNSO) explored the establishment of standardized methods for the collaborative exchange of statistical data. This would involve developing a framework that defines common data formats and definitions. Such a framework would enable computers to automatically exchange data, which would streamline data integration and improve the efficiency of data utilization.

## Data Governance for Government Agencies

In 2019, the Government of Thailand passed the Digitalization of Public Administration and Services Delivery Act, B.E.2562 (2019), which mandates that government agencies establish data governance frameworks. This legislation was enacted to enhance digital management practices, including provisions concerning the exchange of government data, as per Section 8(3) and Section 12(1) of the act.

The TNSO took the lead in developing a data governance framework at the agency level, known as the TNSO Data Governance Framework. This framework was officially implemented on 2 July 2021.

Moreover, the TNSO has embraced international standards, such as the Data Documentation Initiative, the Generic Statistical Business Process Model, and Statistical Data and Metadata eXchange (SDMX), to enhance organizational compatibility in managing statistical data. These standards help to improve data quality, streamline data processes, and facilitate collaboration between agencies.

In addition, the Government of Thailand commissioned the TNSO to develop a central data repository for government agencies. This repository, called the Government Data Catalog, allows agencies to share data in a standardized format, making it easier for users to integrate and analyze data across agencies. While the Government Data Catalog hosts a wealth of datasets across five categories, including statistics, their diverse file formats (CSV, XLSX, PDF, JSON, API) posed a significant challenge for data integration. This stemmed from a lack of standardization across agencies, with varying data storage formats, definitions, and statistical methods hindering seamless collaboration.

To bridge this gap and foster seamless collaboration, the TNSO strategically embraced the SDMX standard. By establishing statistical data structures aligned with SDMX, and publishing them through the Statistics Sharing Hub, the TNSO has orchestrated a remarkable transformation. Data exchange between agencies now flows with unprecedented efficiency, convenience, and speed. This has empowered integration of diverse datasets, such as population and economic statistics, for comprehensive reports and insightful data analysis.

## Technical Aspects of the Implementation

Since 2020, the TNSO has been developing technological infrastructure, promoting SDMX knowledge among its personnel, and raising awareness of the importance of making structured statistical data available. The TNSO has chosen to implement two open-source SDMX tools: the .Stat Suite and the Fusion Metadata Registry.

- .Stat Suite is an open-source platform developed by the Statistical Information System Collaboration Community, supported by the Organisation for Economic Co-operation and Development in collaboration with the European Statistical Office (Eurostat). It is designed for the management and provision of datasets in accordance with the SDMX standard at the Statistics Sharing Hub (https://stathub.nso.go.th/).

- Fusion Metadata Registry is an open-source tool supported by the Bank for International Settlements. It is designed for the creation, maintenance, storage, and management of SDMX artefacts, such as codelists, data structure definitions, and dataflows. It can be accessed at https://sdmx.nso.go.th/.

*continued on next page*

Box 1: *continued*

The TNSO initiated the integration and structured management of all statistical data in accordance with the SDMX standard. It has collected and provided key datasets at the Statistics Sharing Hub, categorized into four groups:

- **International indicator datasets:**
  These datasets cover a wide range of topics, including demographics, economics, and the environment. They are used by a variety of stakeholders, including government agencies, businesses, and researchers.

- **Datasets aligned with national strategic goals:**
  These datasets support the Government of Thailand's key priorities, such as economic development and social welfare. They are used to track progress and inform policy decisions.

- **21 statistical sector datasets:**
  These datasets provide information on specific topics such as agriculture, tourism, and health. They are used by government agencies, businesses, and the public.

- **Datasets presented at the provincial level:**
  These datasets provide provincial-level information on topics such as population, income, and education. They are used by government agencies, businesses, and the public.

Beyond integrating data, the TNSO actively fosters a data-driven ecosystem. They achieve this by:

- **Empowering government agencies:**
  The TNSO regularly organizes SDMX training courses for government agencies, equipping them with the knowledge and tools to utilize and exchange data following international standards. This strengthens interagency collaboration and facilitates seamless data sharing.

- **Building nationwide understanding:**
  The TNSO disseminates knowledge on SDMX through various events and courses, including meetings, seminars, and workshops. This raises awareness about the importance of developing data in accordance with international standards, ultimately promoting data readiness for global use and exchange.

The TNSO's progress has been driven by strong partnerships. The office has collaborated with international organizations such as the Asian Development Bank, the Economic and Social Commission for Asia and the Pacific, the International Labour Organization, the Organisation for Economic Co-operation and Development, the United Nations Children's Fund, and the United Nations Statistics Division. Through these collaborations, the TNSO has received invaluable expertise, training, and technical assistance, which have been the catalysts for the office's remarkable achievements in data integration and development.

## Benefits of Using Statistical Data and Metadata eXchange

The TNSO cites four key benefits in applying the SDMX standard in its Statistics Sharing Hub.

- **Standardized structure for seamless exchange:**
  Collecting and storing data in accordance with SDMX standards guarantees a standardized structure, enabling quick and effortless data exchange across agencies and organizations.

- **Enhanced accessibility and value creation:**
  Adhering to a standardized data structure and utilizing common tools, both domestically and globally, expands data usage across diverse sectors. This promotes convenient and efficient data utilization, reducing time spent on data cleansing.

- **Automated synchronization for accuracy:**
  SDMX programming automates data updates and retrievals, following strict update schedules. This ensures the seamless display of data that are always identical to the corresponding source data.

- **Clear understanding and interoperability:**
  Metadata, codelists, definitions, and classifications managed in compliance with SDMX guidelines foster clear understanding and seamless interoperability among diverse data systems and sources.

## Box 2: Enhancing Data Dissemination in Maldives

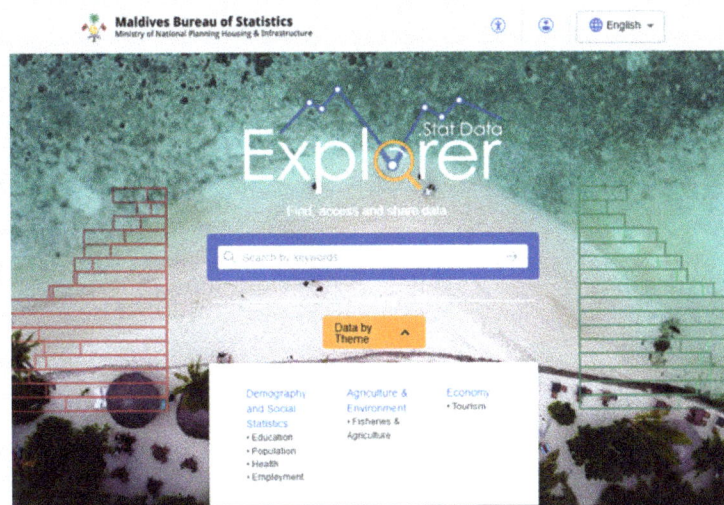

Note: Maldives launched its data portal (https://data.statisticsmaldives.gov.mv) in October 2023 using the .Stat Suite platform and the Statistical Data and Metadata eXchange standard.
Source: Screenshot of the Maldives Bureau of Statistics data portal taken from https://data.statisticsmaldives.gov.mv/.

### Background

In partnership with the United Nations Economic and Social Commission for Asia and the Pacific, the Maldives Bureau of Statistics (MBS) established an initiative aimed at modernizing the data dissemination process in the production of its annual statistical yearbook. The primary objective was to transition from gathering data and publishing in Microsoft Excel format to an advanced interactive data portal using the .Stat Suite. The project not only set out to improve accessibility but also to enhance data efficiency and quality through a more structured and template-driven approach for data collection based on the Statistical Data and Metadata eXchange (SDMX) standard.

### The Role of Statistical Data and Metadata eXchange

By adhering to the SDMX standard, the MBS is ensuring consistency in data structuring, facilitating seamless interoperability and comparability across different datasets. Leveraging SDMX-compliant tools such as the .Stat Suite offers several advantages, including standardized dissemination and better user accessibility for government agencies, researchers, international organizations, and the general public.

Furthermore, the SDMX system enables automated data updates, ensuring the disseminated information remains current and relevant. This feature is particularly crucial for the MBS because it releases new data or updates existing data periodically. Additionally, the SDMX standard provides streamlined data exchange, with international organizations such as the United Nations, the International Monetary Fund, and the World Bank promoting efficient collaboration and data sharing.

### Addressing Challenges in the National Statistical System

One significant challenge faced by the MBS project was that different government agencies in Maldives often collect data in disparate formats, hindering the seamless sharing and comparison of data across agencies and international organizations. The lack of standardized metadata further complicated matters, making it challenging to provide comprehensive descriptions of the data, including definitions, units of measurement, and sources.

The project also encountered constraints related to staff capacity, since the MBS is a relatively small office with only about 50 staff members. The project coincided with other significant tasks, such as the 2022 census, requiring careful resource management and coordination to achieve the objectives.

*continued on next page*

Box 2: *continued*

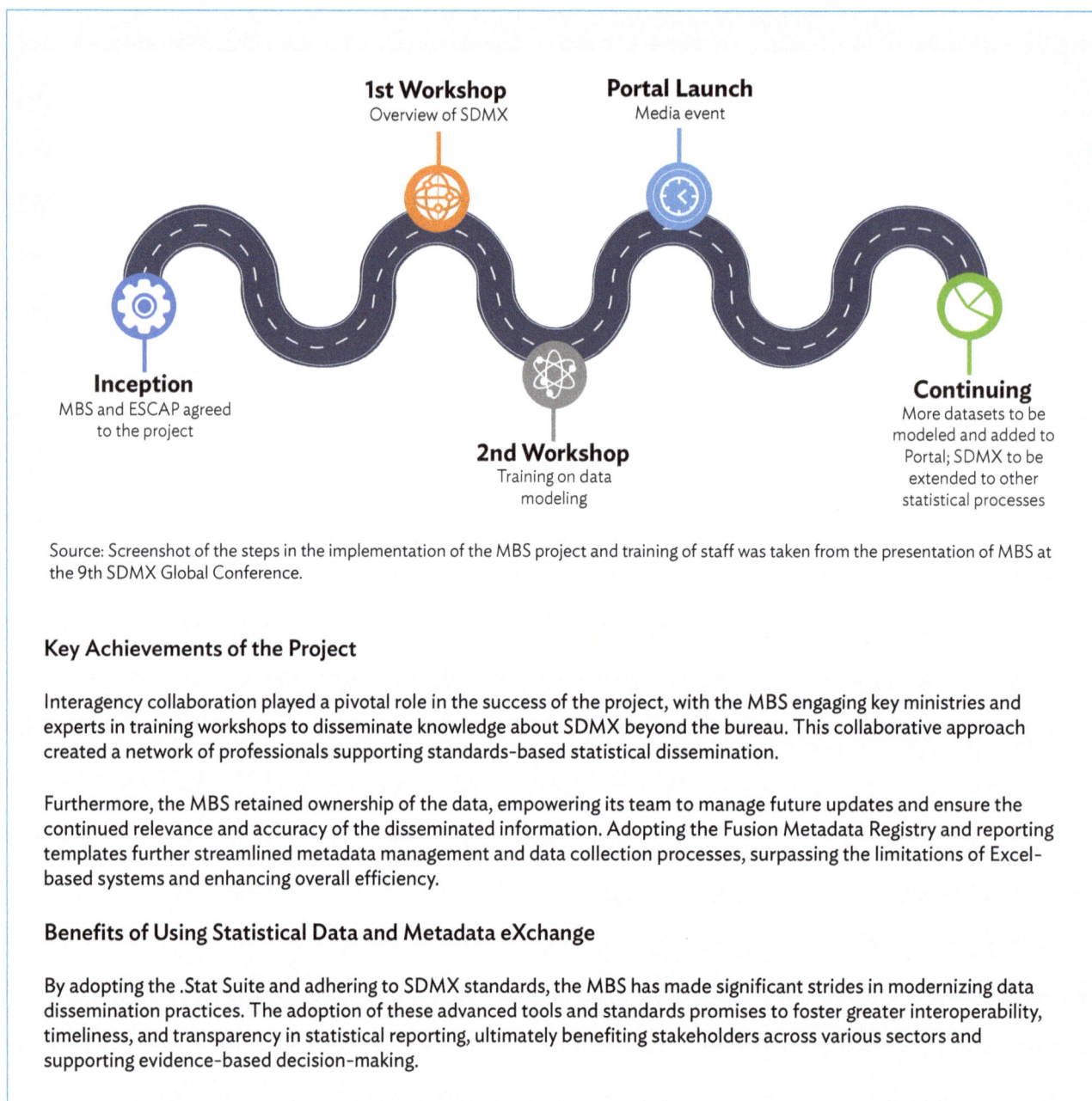

**1st Workshop**
Overview of SDMX

**Portal Launch**
Media event

**Inception**
MBS and ESCAP agreed
to the project

**2nd Workshop**
Training on data
modeling

**Continuing**
More datasets to be
modeled and added to
Portal; SDMX to be
extended to other
statistical processes

Source: Screenshot of the steps in the implementation of the MBS project and training of staff was taken from the presentation of MBS at the 9th SDMX Global Conference.

### Key Achievements of the Project

Interagency collaboration played a pivotal role in the success of the project, with the MBS engaging key ministries and experts in training workshops to disseminate knowledge about SDMX beyond the bureau. This collaborative approach created a network of professionals supporting standards-based statistical dissemination.

Furthermore, the MBS retained ownership of the data, empowering its team to manage future updates and ensure the continued relevance and accuracy of the disseminated information. Adopting the Fusion Metadata Registry and reporting templates further streamlined metadata management and data collection processes, surpassing the limitations of Excel-based systems and enhancing overall efficiency.

### Benefits of Using Statistical Data and Metadata eXchange

By adopting the .Stat Suite and adhering to SDMX standards, the MBS has made significant strides in modernizing data dissemination practices. The adoption of these advanced tools and standards promises to foster greater interoperability, timeliness, and transparency in statistical reporting, ultimately benefiting stakeholders across various sectors and supporting evidence-based decision-making.

## Box 3: Unlocking Data Access Across the Pacific

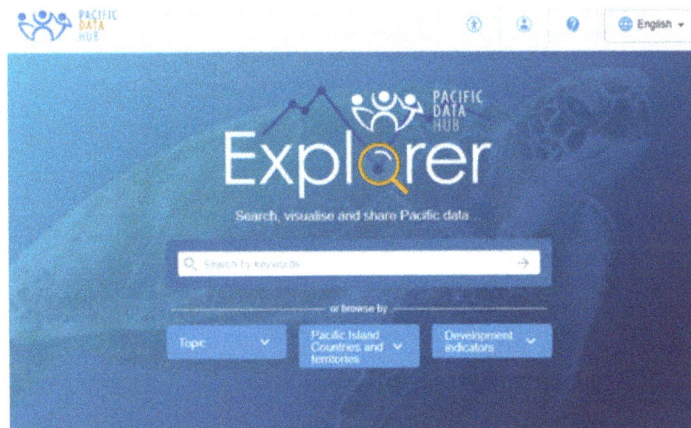

Note: The Pacific Data Hub, developed with support from the Pacific Community, is accessible via https://stats.pacificdata.org/.
Source: Screenshot of the Pacific Data hub taken from https://stats.pacificdata.org/.

### Background

The Pacific Data Hub, established by the Pacific Community, is enhancing data accessibility and promoting evidence-based policymaking for sustainable development across the subregion. One significant initiative in this endeavor is the adoption of the Statistical Data and Metadata eXchange (SDMX) standard, coupled with the use of the .Stat Suite, to unlock data access and support comprehensive data management.

### *Enhancing Data Sharing and Access across the Pacific*

By adhering to the SDMX standard, the Pacific Data Hub ensures interoperability and the seamless exchange of data among various stakeholders, including government agencies, international organizations, and research institutions. This standardization streamlines data sharing and integration processes, enabling more efficient use of statistics for informed policymaking and development planning.

Furthermore, use of the .Stat Suite complements the SDMX standard by providing a robust platform for finding, analyzing, and visualizing data. The suite represents a paradigm shift in access to statistics, offering users an intuitive browser-based application to explore a vast array of indicators. Its dynamic and flexible interface allows users to extract information effortlessly from extensive data tables.

### Support for the Sustainable Development Goals

The Pacific Data Hub plays a key role in supporting progress towards the Sustainable Development Goals (SDGs) in the Pacific subregion. Data collected from various sources, including Pacific economies and international agencies, are centralized on the hub. Using the data hub, stakeholders can access an SDG dashboard tailored to the Pacific context. This dashboard offers dynamic visualizations of key indicators, enabling policymakers and organizations to track regional progress toward achieving the SDGs and to make informed decisions for sustainable development in the subregion.

*continued on next page*

Box 3: *continued*

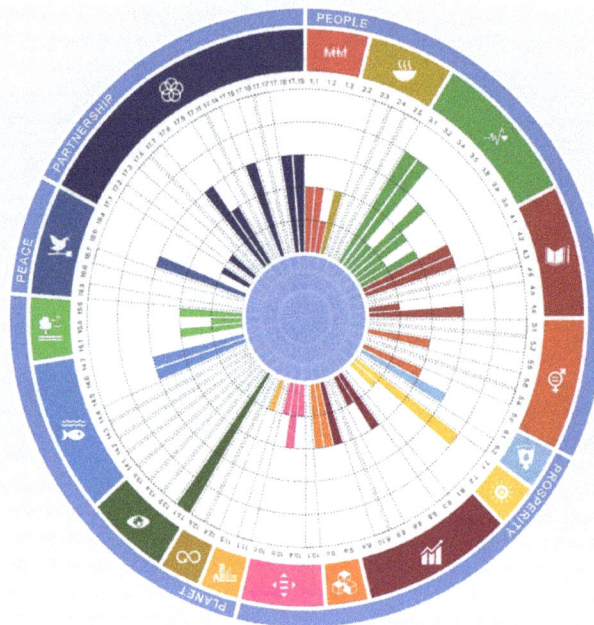

Note: The SDG dashboard under the Pacific Data Hub is accessible via
https://pacificdata.org/dashboard/17-goals-transform-pacific.
Source: Screenshot of the SDG dashboard under the Pacific Data Hub
taken from https://pacificdata.org/dashboard/17-goals-transform-pacific.

### Advancing Modern Statistical Systems and Practices

Through initiatives such as the SDMX-compliant data hub, the Pacific Community is leading the modernization of
the subregion's statistics system by supporting the adoption of state-of-the-art data management practices. Through
partnerships with key stakeholders such as the Asian Development Bank and the United Nations Economic and Social
Commission for Asia and the Pacific, the Pacific Community is helping small island economies and territories to navigate
the implementation of hi-tech statistical systems and is empowering these economies to make informed policy decisions
that will foster sustainable development and socioeconomic progress in the Pacific.

# 8 ADB ECONOMY ASSESSMENT SURVEY

## 8.1 Overview of the Survey

In collaboration with the United Nations Statistics Division, the Economic and Social Commission for Asia and the Pacific, and the Statistical Institute for Asia and the Pacific, ADB conducted a statistical assessment survey from 31 May to 9 July 2021. The primary objective of this survey was to gather insights from national statistics offices (NSOs) across the bank's members in the Asia and Pacific region, focusing on their familiarity with the Statistical Data and Metadata eXchange (SDMX) standard and the potential for its integration within their national statistical system (NSS). The survey also aimed to identify the specific priorities and requirements of these NSOs in relation to the implementation of SDMX. Notably, 31 of ADB's 49 members in Asia and the Pacific participated in the survey.

It is worth mentioning here that surveys related to SDMX had been conducted before the ADB economy assessment survey of 2021. A notable example is the 2016 report published by the Irving Fisher Committee on Central Bank Statistics (IFC), which specifically examined the utilization and interest in SDMX among central banks (Ehrmann et al. 2016). In 2019, the Organisation for Economic Co-operation and Development conducted the SDMX Global Survey, which was completed by NSOs, central banks, international organizations, and other government agencies. This 2019 survey aimed to assess the level of acceptance, challenges faced, and implementation plans for SDMX while also intending to capture some needs and priorities within the global statistics community (SDMX 2019).

## 8.2 Results of the ADB Survey

The results below offer a collective understanding of the interests of NSOs in utilizing SDMX, the varying levels of interest based on SDMX usage, the specific areas of focus in SDMX implementation, the perceived benefits associated with SDMX, and challenges encountered in adopting SDMX.

Table 3 presents the level of interest NSOs had in using SDMX. The results are grouped by subregion as well as a singular category grouping the developed economies. The table provides insights on whether respondents were using SDMX in their NSS; expressed interest in using SDMX; or had no plans to incorporate SDMX. It indicates that the majority of NSOs in all subregions had either adopted or were interested in using SDMX.

Both the earlier IFC report and the survey conducted by ADB highlighted the prevalence of SDMX implementation, with the IFC report indicating that 64% of central banks surveyed in 2019 were using SDMX, while the ADB survey found that 48.4% of NSOs were using SDMX in 2021. Additionally, both surveys show that a significant percentage of respondents planned to use SDMX.

Table 3: Interest in Using Statistical Data and Metadata eXchange, by Subregion

| Region | Uses it now | Interested to use it | Does not plan to use it |
|---|---|---|---|
| Central and West Asia | 3 | 2 | 1 |
| East Asia | 2 | 1 | 1 |
| South Asia | 1 | 2 | 2 |
| Southeast Asia | 4 | 3 | 2 |
| Pacific | 3 | 2 | 0 |
| Developed Economies | 2 | 0 | 0 |
| All Regions | 15 | 10 | 6 |

Source: Asian Development Bank estimates based on the results of the 2021 Statistical Data and Metadata eXchange economy assessment survey.

The IFC report and the ADB survey also showed comparable percentages of respondents who did not plan to use SDMX. The IFC report found that 23% of central banks surveyed in 2019 did not plan to use the standard, while the ADB survey in 2021 found that 19.4% NSOs did not plan to use it. The challenges of implementing SDMX are described in more detail in Figure 16.

Table 4 provides an overview of the status of SDMX implementation by NSOs across different statistical domains. It highlights whether the implementation is already operational, undergoing pilot testing, or still in the planning phase. Note that an NSO may already have implemented SDMX in one or more statistical domains, but is still planning or piloting SDMX adoption for other domains.

Table 4: Implementation Status of Statistical Data and Metadata eXchange, by Statistical Domain

| Domain | Status | | |
|---|---|---|---|
| | Planned | Pilot | Production |
| SNA | 7 | 1 | 10 |
| BOP | 6 | 1 | 9 |
| SDDS Plus/e-GDDS | 8 | 0 | 9 |
| Labor | 7 | 2 | 8 |
| Prices | 7 | 2 | 8 |
| GFS | 5 | 1 | 8 |
| SDGs | 6 | 2 | 8 |
| IMTS | 6 | 3 | 7 |
| Education | 8 | 0 | 6 |
| FDI | 8 | 1 | 4 |
| SEEA | 9 | 0 | 1 |

BOP = Balance of Payments, e-GDDS = Enhanced General Data Dissemination System, FDI = Foreign Direct Investment, GFS = Government Finance Statistics, IMTS = International Merchandise Trade Statistics, SDDS Plus = Special Data Dissemination Standard Plus, SDGs = Sustainable Development Goals, SEEA = System of Environmental Economic Accounting, SNA = System of National Accounts.
Source: Asian Development Bank estimates based on the results of the 2021 Statistical Data and Metadata eXchange economy assessment survey.

In terms of SDMX implementations already in production, Table 4 shows that the leading statistical domains were the System of National Accounts (SNA), balance of payments (BOP), and Special Data Dissemination Standard (SDDS) Plus or the Enhanced General Data Dissemination System (e-GDDS). Notably, SNA and BOP were among the pioneering global Data Structure Definitions (DSDs) released in 2013. Following closely are domains such as Labor, Prices, government finance statistics (GFS), and Sustainable Development Goals (SDGs).

Table 5 provides a breakdown of NSOs level of interest among NSOs in adopting SDMX, categorized according to various types of usage.

### Table 5: Interest in Using Statistical Data and Metadata eXchange, by Type of Usage

| Purpose | Total | Highly Interested | Interested | Slightly Interested | Not Interested |
|---|---|---|---|---|---|
| Data dissemination | 31 | 14 | 15 | 1 | 1 |
| Data reporting | 31 | 10 | 19 | 1 | 1 |
| Visualize data | 31 | 14 | 14 | 3 | 0 |
| Transform/convert between SDMX formats | 31 | 11 | 14 | 6 | 0 |
| Internal data management | 31 | 11 | 15 | 2 | 3 |
| Validate data files | 31 | 10 | 13 | 6 | 2 |
| Collect data | 31 | 10 | 14 | 3 | 4 |
| Map/transcode between DSDs | 31 | 9 | 16 | 4 | 2 |
| Average | 31 | 11 | 15 | 3 | 2 |

DSD = data structure definition, SDMX = Statistical Data and Metadata eXchange.
Source: Asian Development Bank estimates based on the results of the 2021 Statistical Data and Metadata eXchange economy assessment survey.

Table 5 reveals that 29 or the 31 participating NSOs had a keen interest in using SDMX for data dissemination and data reporting, with 28 of 31 NSOs also expressing genuine interest in using SDMX data visualization. Among the other five types of usage, 23–26 of the 31 participating NSOs were inclined toward utilizing SDMX for these purposes. Across all purposes, the average number of NSOs either highly interested or interested in adopting SDMX was 26, indicating a general enthusiasm for the standard and signaling significant potential for enhancing the efficiency and effectiveness of statistical data exchange and management.

Figure 15 outlines a list of the perceived benefits of SDMX as ranked by NSOs. These benefits are categorized according to the number of participating NSOs, grouped by subregion, who rated each as their top-ranked benefit.

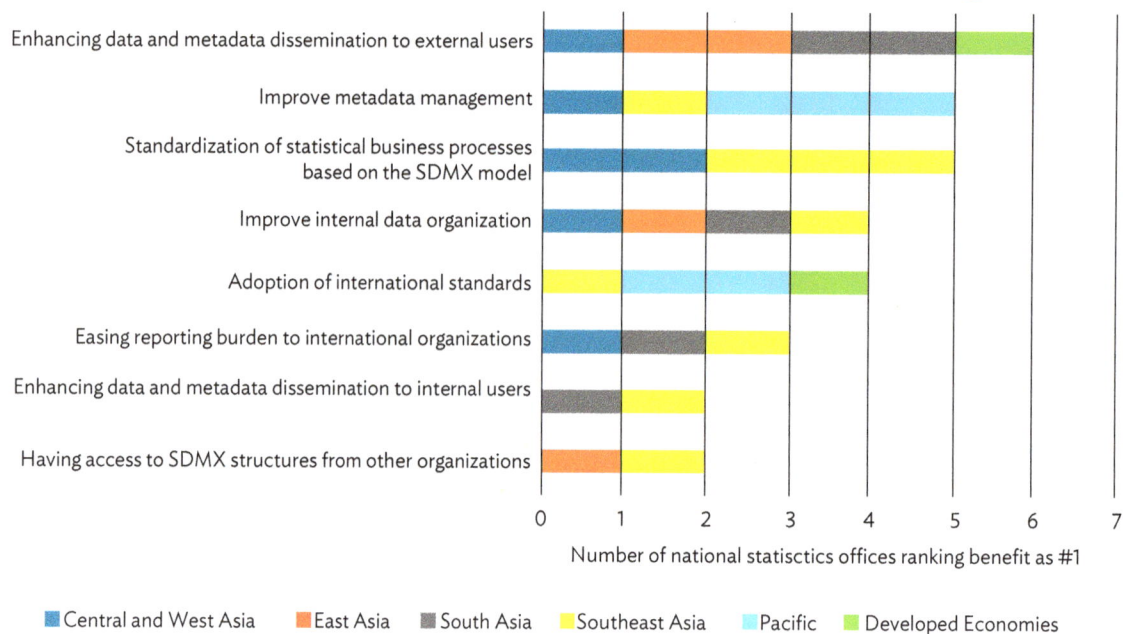

Figure 15: Perceived Benefits of Statistical Data and Metadata eXchange

SDMX = Statistical Data and Metadata eXchange.
Source: Asian Development Bank estimates based on the results of the 2021 Statistical Data and Metadata eXchange economy assessment survey.

Figure 15 indicates a substantial emphasis on adopting and utilizing the SDMX standard to improve data and metadata dissemination to external users, except among participating NSOs in Southeast Asia and the Pacific. This benefit is closely followed by a focus on enhancing metadata management, with significant interest from NSOs in the Pacific. Standardizing statistical business processes is perceived to be beneficial for the NSOs in Southeast Asia and Central and West Asia. The least-emphasized benefits of SDMX across various subregions were enhancing data and metadata dissemination within the organization and having access to SDMX structures from other organizations.

Figure 16 shows the survey results related to challenges faced by NSOs in implementing SDMX into their NSS. Again, these perceived challenges are presented in a ranked order, according to the number of NSOs that identified each as their topmost challenge in the SDMX implementation process. The participating NSOs are again grouped by subregion.

The SDMX implementation challenge ranked highest in Figure 16 was a lack of relevant tools and training resources, as nominated by NSOs across all five subregions but not by the developed economies. This challenge was followed by the need to obtain support from subject-matter statisticians, with particular emphasis from the Pacific, Southeast Asia, South Asia, and East Asia. Lack of financial and/or human resources was perceived as a challenge in Central and West Asia, East Asia, South Asia, and in the developed economies. Interestingly, the lowest-ranked implementation challenge was conflict with existing data transmission strategies, suggesting a general preparedness to integrate the SDMX standard into existing practices.

## Figure 16: Perceived Challenges of Implementing Statistical Data and Metadata eXchange

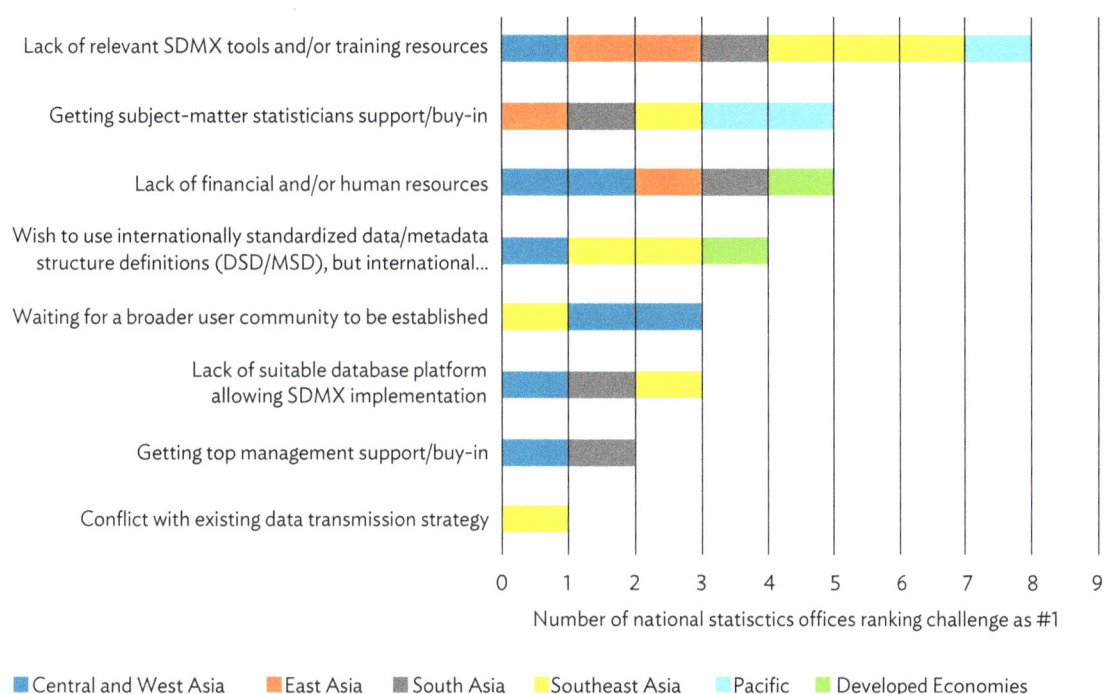

Number of national statisctics offices ranking challenge as #1

■ Central and West Asia  ■ East Asia  ■ South Asia  ■ Southeast Asia  ■ Pacific  ■ Developed Economies

DSD = data structure definition, MSD = metadata structure definition, SDMX = Statistical Data and Metadata eXchange.
Source: Asian Development Bank estimates based on the results of the 2021 Statistical Data and Metadata eXchange economy assessment survey.

# 8.3  Key Takeaways of the Survey

• A strong majority of the participating NSOs across Asia and the Pacific had either used SDMX or expressed a keen interest in integrating SDMX into their NSS.

• The System of National Accounts (SNA) stood out as the top statistical domain for which SDMX implementation has been prioritized.

• Improved data dissemination and data reporting emerged as significant reasons for adoption of the SDMX standard.

• Enhancing data and metadata dissemination was perceived by NSOs across various subregions as the top benefit of SDMX.

• A lack of relevant tools and training resources continues to be the primary constraint to SDMX implementation across all subregions (but not for the developed economies).

In general, most of the participating NSOs from across Asia and the Pacific indicated that they would focus efforts in implementing SDMX on data and metadata dissemination and/or reporting. This highlights the importance NSOs place on ensuring that their statistical data and related metadata are efficiently and effectively disseminated to internal and external users. Standardizing data exchange through SDMX can contribute to more seamless and timely data sharing, supporting evidence-based decision-making.

# 9 ELECTRONIC COURSES SUPPORTING STATISTICAL DATA AND METADATA EXCHANGE

Findings from ADB's 2021 Statistical Data and Metadata eXchange (SDMX) Economy Assessment Survey have played an essential role in the development of advanced capacity-building resources for SDMX.

Because the primary challenge national statistics offices (NSOs) faced in implementing SDMX was a lack of relevant training tools, ADB has collaborated with the United Nations Statistics Division, the Economic and Social Commission for Asia and the Pacific, and the Statistical Institute for Asia and the Pacific to create and conduct e-learning training programs that address specific needs and challenges identified in the survey.

The e-learning programs are tailored to different levels of expertise, with participants able to undertake either or both the SDMX Foundation Course[12] and the SDMX Tools Course.[13]

## 9.1 Foundation Course

### 9.1.1 Objective, Structure, and Development of the Foundation Course

The primary objective of the Foundation Course is to provide education on the fundamentals of SDMX for managers, supervisors, statisticians, IT experts, and other personnel from NSOs, line ministries, regional and international organizations, and other entities engaged in statistical data and metadata management.

The course is divided into two modules. Module 1 is a high-level introduction tailored to decision makers. It delves into the rationale and advantages of SDMX, and how it can fit into and improve the statistical business process. Module 2 is a more in-depth exploration of the foundational concepts of SDMX and is intended for practitioners and those considering implementing SDMX in the future. It provides detailed insights on the SDMX Information Model and the SDMX Content-Oriented Guidelines along with SDMX infrastructure and IT tools.

As an e-learning platform, the SDMX Foundation Course can be completed remotely, offering self-paced flexibility of learning and travel cost savings. The course structure is shown in Table 6.

---

[12] Access the SDMX Foundation e-learning course via https://elearn.adb.org/course/view.php?id=486. Access the user guide via https://elearn.adb.org/mod/book/view.php?id=10142&chapterid=2138.

[13] Access the SDMX Tools e-learning course via https://elearn.adb.org/course/view.php?id=520.

Table 6: Structure of the Foundation Course

| Module | Submodule | Outline |
|---|---|---|
| 1 | | Module 1 provides participants with a high-level introduction to what SDMX is and how it can fit into and improve statistical business processes |
| | 1.1 –1.4 | Introduce learners to quick introduction of SDMX:<br>• Welcome to the SDMX Foundation Course<br>• Navigating the course<br>• Quick introduction to SDMX<br>• Sample presentation of NSOs data processes |
| | 1.5 – 1.8 | Discuss with learners the various issues, challenges, and business case:<br>• Car production analogy<br>• Issues and challenges in National Statistical Systems<br>• Business case for SDMX<br>• Agency experience: Australian Bureau of Statistics, National Statistical Office of Thailand, Pacific Community |
| | 1.9 – 1.13 | Show the various sample implementations:<br>• How SDMX can be applied in an organizations<br>• Limitations of SDMX<br>• Evolution of SDMX |
| 2 | | Module 2 introduces participants to basic information on how SDMX works: the information model, content-oriented guidelines, and SDMX infrastructure and IT tools. This module also provides a recap and more information on things to consider when preparing for SDMX implementation. |
| | 2.1 – 2.2 | Present to the learners the introduction of Module 2 and overview of SDMX components |
| | 2.3 – 2.7 | Detailed discussion of the SDMX Information Model:<br>• Introduction to Data Modeling<br>• Key parts of SDMX Information Model such as Data Structure Definition, Concept Scheme, Codelists, etc. |
| | 2.8 – 2.10 | Detailed discussion of the SDMX Content-Oriented Guidelines such as cross-domain concepts, codelists, statistical subject-matter domains, and SDMX glossary |
| | 2.11 – 2.12 | Detailed discussion of the SDMX Infrastructure and IT Tools |
| | 2.13 – 2.14 | Show the various steps of SDMX implementation, key takeaways, and next steps for SDMX |

IT = information technology, NSOs = national statistics offices, SDMX = Statistical Data and Metadata eXchange.
Source: Asian Development Bank construction based on information from https://elearn.adb.org/course/view.php?id=486.

The timeline for creating and conducting the first intake of the SDMX Foundation Course is shown in Figure 17.

## 9.1.2 Demographics of Foundation Course Graduates

Table 7 provides an overview of the enrollment status and success rates for the first intake of the SDMX Foundation Course. Of the 692 individuals who enrolled, 606 students actually started the course and 433 successfully passed the course in the first iteration during May 2022.

After the reopening of the SDMX Foundation Course during February 2023, an additional 127 learners successfully passed the course. This brought the count of certified SDMX Foundation graduates to 560. The gender distribution among people who successfully completed and passed the course shows a balanced representation, with 50% of graduates being women, 49% men, and 1% electing to not nominate a gender.

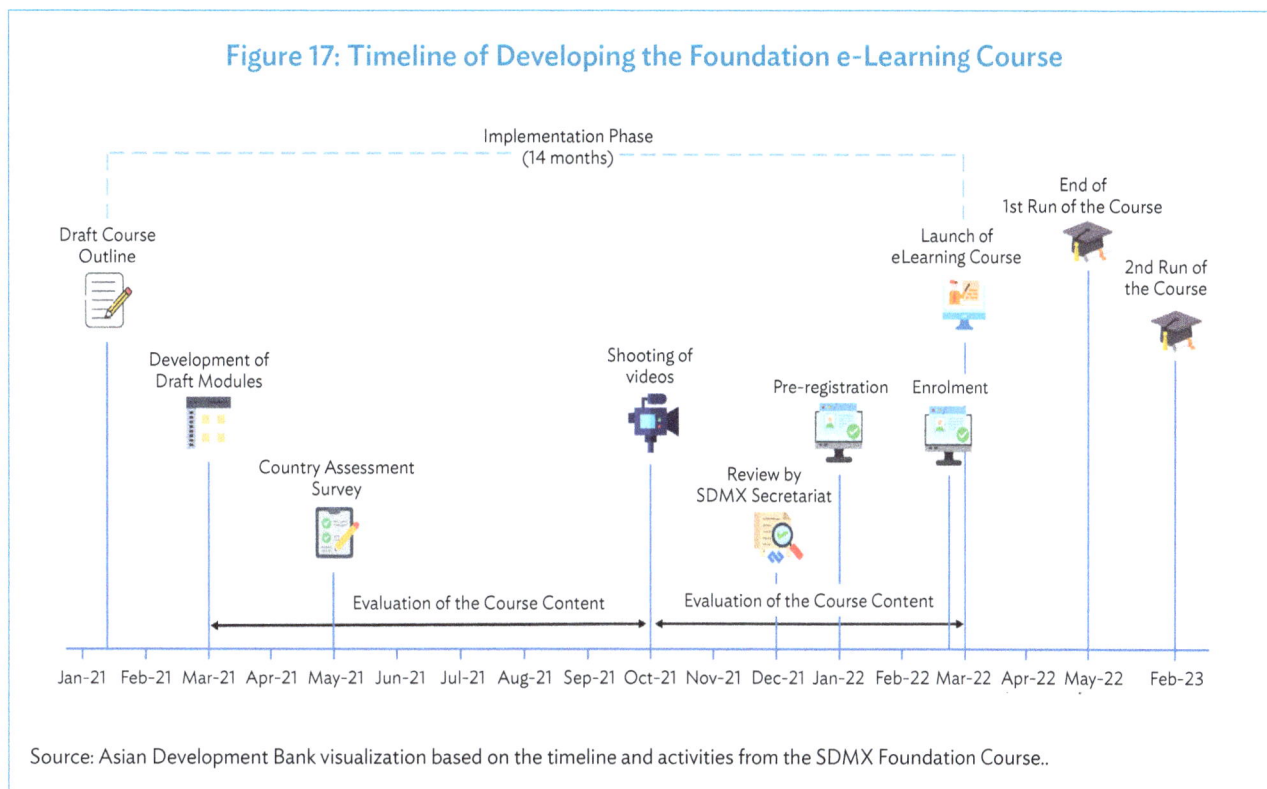

Figure 17: Timeline of Developing the Foundation e-Learning Course

Source: Asian Development Bank visualization based on the timeline and activities from the SDMX Foundation Course..

Table 7: Completion Rate of the Foundation Course

| Status | Frequency of Learners |
|---|---|
| Enrolled (first iteration) | 692 |
| Enrolled and started (first iteration) | 606 |
| Enrolled but did not start (first iteration) | 86 |
| Passed (first iteration) | 433 |
| Passed (second iteration) | 127 |
| Total passed* | 560 |

* Total number of learners who completed Module 1 and Module 2 of the SDMX Foundation Course and had a cumulative grade in all quizzes of at least 70%.
Source: Asian Development Bank estimates based on the evaluation report of the SDMX Foundation Course.

Of the participants who passed the SDMX Foundation Course, 77.7% were below the age of 44 years, with 4.3% of graduates not specifying their age (Table 8).

Of the successful participants in the first intake of the SDMX Foundation Course, 60% were from ADB member economies and 36% were from other economies, with 4% unspecified (Figure 18). Geographic analysis reveals that 52% of those who passed the course were from East Asia, followed by 26% from Southeast Asia, and 12% from South Asia. Furthermore, among graduates from ADB members, NSO staff from 23 economies completed and passed the Foundation Course. It is also worth noting that staff from six out of the 10 NSOs that expressed interest in using SDMX in the ADB economy assessment survey are among the graduates, as well as staff from three NSOs that indicated no interest in SDMX.

### Table 8: Age Distribution of Foundation Course Graduates

| Age Group (years) | Number of Learners | % |
|---|---|---|
| 18–24 | 28 | 5.0 |
| 25–34 | 240 | 42.9 |
| 35–44 | 167 | 29.8 |
| 45–54 | 75 | 13.4 |
| 55–64 | 26 | 4.6 |
| Not Specified | 24 | 4.3 |
| Total | 560 | 100 |

Source: Asian Development Bank estimates based on evaluation report of the SDMX Foundation Course.

### Figure 18: Origins of Foundation Course Graduates

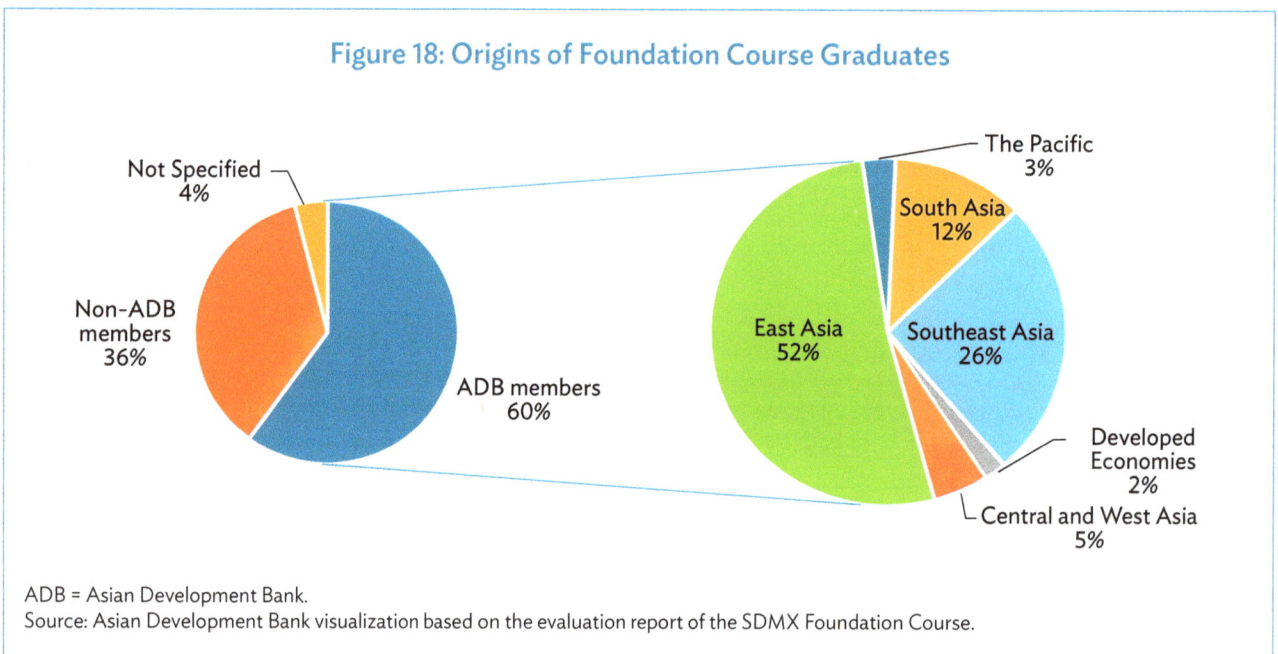

ADB = Asian Development Bank.
Source: Asian Development Bank visualization based on the evaluation report of the SDMX Foundation Course.

In terms of affiliation or organization, 62.7% of SDMX Foundation Course graduates in the first cohort were from NSOs, while other government organizations accounted for 11.8% (Table 9). The private sector provided the lowest percentage of graduates, with only 1.4%.

Filtering by job classification, 62.7% of those who completed and passed the first SDMX Foundation Course were statisticians, economists, or held other data-related positions. This was followed by IT personnel and people in managerial or supervisorial positions, who collectively represented 30.7% of all the successful participants (Table 10).

## Table 9: Affiliation or Organization of Foundation Course Graduates

| Type of Agency | Number of Graduates | % |
|---|---|---|
| National statistics office | 351 | 62.7 |
| International/regional organization | 31 | 5.5 |
| Central bank | 39 | 7.0 |
| Private company | 8 | 1.4 |
| Other government agency | 66 | 11.8 |
| Others | 37 | 6.6 |
| Academe | 2 | 0.4 |
| Independent/Self-employed | 2 | 0.4 |
| Not Specified | 24 | 4.2 |
| Total | 560 | 100 |

Source: Asian Development Bank estimates based on the evaluation report of the SDMX Foundation Course.

## Table 10: Job Classification of Foundation Course Graduates

| Type of Profession | Number of Graduates | % |
|---|---|---|
| Statistician, economist, and data-related job | 351 | 62.7 |
| Information technology-related job | 74 | 13.2 |
| Managerial or supervisorial position | 98 | 17.5 |
| Others | 13 | 2.3 |
| Not Specified | 24 | 4.3 |
| Total | 560 | 100 |

Source: Asian Development Bank estimates based on the evaluation report of the SDMX Foundation Course.

## 9.1.3  Foundation Course Evaluation by Graduates

Table 11 presents key feedback from people who successfully passed the first intake of the SDMX Foundation Course. An overwhelming majority of the graduates gave high ratings (excellent or good) for all specific aspects of the course: (i) clarity of topics, 95%; (ii) presentation style or delivery of the course lecturer, 95.5%; and (iii) relevance of the content to their work, 88.9%. Similarly high ratings were given for both Module 1 and Module 2.

## 9.1.4. Foundation Course Results Summary

Overall, the SDMX Foundation Course is regarded as being extremely successful, both in terms of completion rates and graduate reviews. It is seen as a comprehensive and practical response to the lack of SDMX training resources identified as a primary challenge for NSOs in ADB's 2021 economy assessment survey.The course has proven to be instrumental in imparting essential foundational knowledge on SDMX, particularly for participants from data-producing agencies such as NSOs and central banks.

## Table 11: Graduate Evaluation of the Foundation Course

| Evaluation | Number of Graduates | % |
|---|---|---|
| **Topics were clearly explained** | | |
| Excellent | 395 | 70.5 |
| Good | 137 | 24.5 |
| Satisfactory | 18 | 3.2 |
| Fair | 3 | 0.5 |
| Poor | 0 | 0.0 |
| No Response | 7 | 1.3 |
| **Presentation style/delivery of course lecturer** | | |
| Excellent | 401 | 71.6 |
| Good | 134 | 23.9 |
| Satisfactory | 15 | 2.7 |
| Fair | 1 | 0.2 |
| Poor | 1 | 0.2 |
| No Response | 8 | 1.4 |
| **Relevance of content to your work** | | |
| Excellent | 326 | 58.2 |
| Good | 172 | 30.7 |
| Satisfactory | 47 | 8.4 |
| Fair | 8 | 1.4 |
| Poor | 0 | 0.0 |
| No Response | 7 | 1.3 |
| **Overall Rating of Module 1[a]** | | |
| Excellent | 396 | 70.7 |
| Good | 141 | 25.2 |
| Satisfactory | 13 | 2.3 |
| Fair | 3 | 0.5 |
| Poor | 0 | 0.0 |
| No Response | 7 | 1.3 |
| **Overall Rating of Module 2[b]** | | |
| Excellent | 394 | 70.4 |
| Good | 140 | 25.0 |
| Satisfactory | 17 | 3.0 |
| Fair | 2 | 0.4 |
| Poor | 0 | 0.0 |
| No Response | 7 | 1.3 |

[a] Module 1 covers the introduction to Statistical Data and Metadata eXchange (SDMX), discusses various issues and challenges, presents the business case for SDMX, and provides examples of different SDMX implementations.
[b] Module 2 provides an overview of the SDMX components, a detailed discussion of the SDMX information model, content-oriented guidelines, and SDMX IT tools, and presents the various steps for SDMX.
Source: Asian Development Bank estimates based on the evaluation report of the SDMX Foundation Course.

Most of the first cohort of SDMX Foundation Course graduates fell within the working-age demographic and held occupations as statisticians, economists, data-related professionals, IT personnel, or were employed in managerial or supervisory positions.

Although the majority of first intake graduates were from ADB members, it is noteworthy that the SDMX Foundation Course has also attracted individuals from other economies.

It is also worth noting that three out of the six NSOs who reported that they did not plan to use SDMX, as outlined in Table 3, have had staff enroll in and pass the course. This indicates a growing interest in SDMX since the ADB survey in 2021.

Moreover, while the majority of first intake graduates were from ADB members, it is noteworthy that the SDMX Foundation Course has also attracted individuals from other economies.

Important note: The online SDMX Foundation Course remains open for enrollments.[14] The course offers remote and cost-efficient learning, with self-pacing to accommodate the participant's preferred schedule.

# 9.2  Tools Course

## 9.2.1  Objective, Structure, and Development of the Tools Course

The SDMX Tools Course was developed to familiarize participants with more detailed technological aspects of the standard. This course is tailored to statisticians and IT experts, but also has appeal for professionals from NSOs, line ministries, regional and international organizations, and entities engaged in statistical data and metadata management.

The SDMX Tools Course offers a comprehensive assessment of SDMX tools, presenting common scenarios and usage applications. It provides an in-depth exploration of the features and practical demonstrations associated with three key SDMX tools: the SDMX Constructor, the Fusion Metadata Registry, and the SDMX Converter.

The structure of the SDMX Foundation Course is shown in Table 12. Module 1 is an outline of the various SDMX tools, giving participants a good foundation for understanding how the three tools presented in detail under Module 2 fit into the broader SDMX scheme.

---

14   Access enrollment details via https://elearn.adb.org/course/view.php?id=486.

### Table 12: Structure of the Tools Course

| Module | Submodule | Outline |
|---|---|---|
| 1 | | Module 1 provides participants with an outline of the different SDMX tools and their purposes, along with various usage applications and comparisons to help participants decide which tools to use. |
| | 1.1 –1.2 | Introduce learners to quick introduction of SDMX:<br>• Module 1 objectives, expected learning outcomes, and target audience<br>• Navigating the course<br>• Background of SDMX |
| | 1.3 | Discuss with learners the various SDMX IT Tools for different usage:<br>• Data file converters<br>• Database mapping suite<br>• Toolkits/Suites<br>• Structural metadata maintenance tools<br>• SDMX Registry<br>• Visualization and analysis tools and connectors<br>• Validation tools |
| | 1.4 | Present to the learners the various SDMX use cases such as data reporting, data collection, and data dissemination |
| 2 | | Module 2 introduces participants to the step-by-step guide on how to use the select SDMX tools, including installation, discussion of features and functions, and output each tool can produce. |
| | 2.1 | Present to the learners the introduction of Module 2:<br>• Module 2 objectives, expected learning outcomes, and target audience<br>• Quick recap of Module 1<br>• Select scenarios for Module 2<br>• Overview of the selected SDMX tools |
| | 2.2 | Detailed discussion of the SDMX Constructor:<br>• Overview of SDMX Constructor<br>• Installation, features, and functions<br>• Demo on DSD creation and customization<br>• Hands-on exercise on DSD creation |
| | 2.3 | Detailed discussion of the Fusion Metadata Registry:<br>• Overview of FMR<br>• Installation, navigation, features and functions<br>• Demo on exporting SDMX artefacts, SDMX artefact management, using REST API web service, and other features<br>• Hands-on exercise on SDMX Artefact Management |
| | 2.4 | Detailed discussion of the SDMX Converter:<br>• Overview of SDMX Converter<br>• Demo on mapping of Excel datasets, and conversion and validation of data file<br>• Hands-on exercise on data conversion and validation |
| | 2.5 | Show the key takeaways of the SDMX Tools Course |

DSD = data structure definition, FMR = Fusion Metadata Registry, IT = information technology, REST API = representational state transfer application programming interface, SDMX = Statistical Data and Metadata eXchange.
Source: Asian Development Bank construction based on information from https://elearn.adb.org/course/view.php?id=520.

The timeline for creating and conducting the first intake of the SDMX Tools Course is shown in Figure 19.

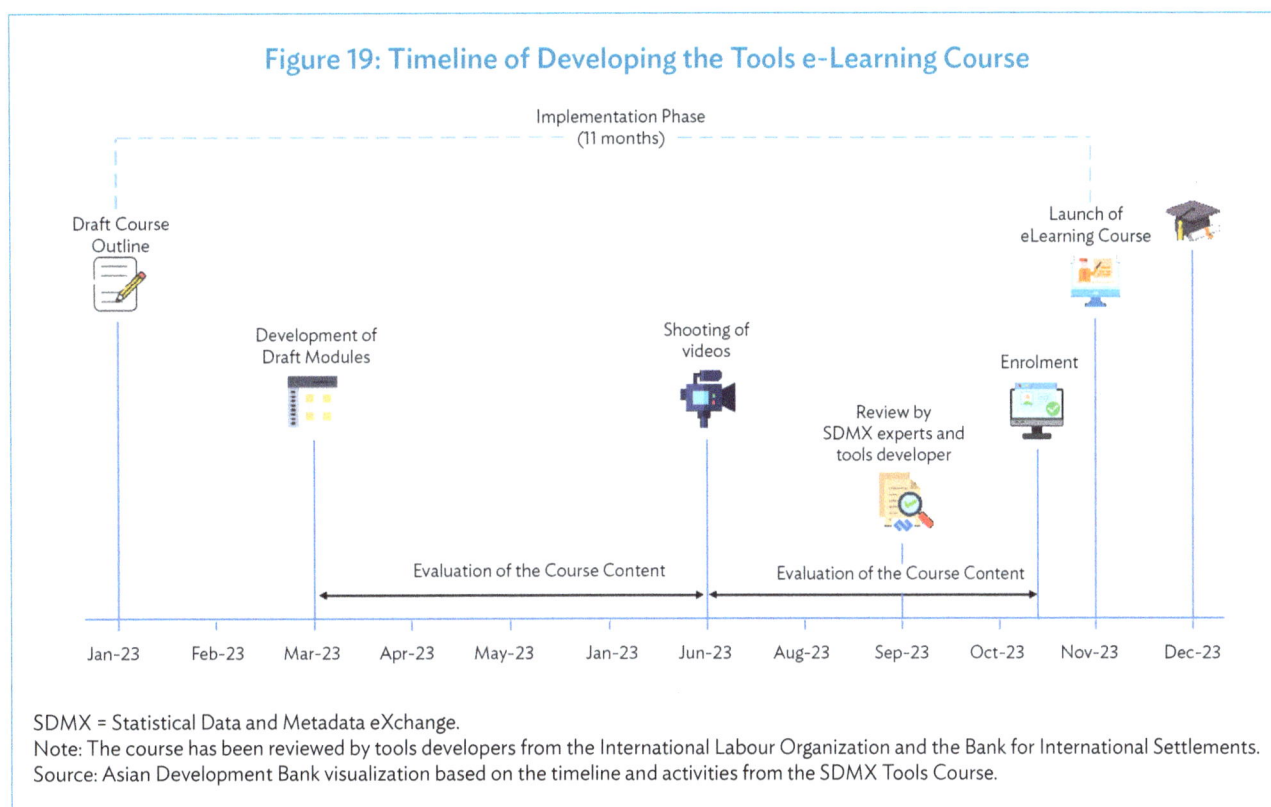

**Figure 19: Timeline of Developing the Tools e-Learning Course**

SDMX = Statistical Data and Metadata eXchange.
Note: The course has been reviewed by tools developers from the International Labour Organization and the Bank for International Settlements.
Source: Asian Development Bank visualization based on the timeline and activities from the SDMX Tools Course.

## 9.2.2 Demographics of Tools Course Graduates

Table 13 provides an overview of the enrollment status and success rates for the first intake of the SDMX Tools Course. The first intake commenced on 15 November 2023 and ran for 5 weeks. There were a total of 276 enrollees of whom only 170 started the course. Among these participants who actively started the course, 66 people successfully completed all the course activities, passed with a cumulative grade of at least 70% across all the quizzes and exercises, and submitted the requirements for the practical exercises on the SDMX Constructor, the Fusion Metadata Registry, and the SDMX Converter.

While the completion rate of 39% for the SDMX Tools Course is reasonable for a specialized online training program, it is markedly lower than the 92.4% achieved in the SDMX Foundation Course. This is likely due to the more difficult nature of the material. In particular, most of the participants who failed to finish the course dropped out when required to do the SDMX Constructer exercise (first tool in Module 2). It should be noted that, during the first intake of the SDMX Tools Course, there were no online tools available for the SDMX Constructor and participants were required to install the SDMX Constructor on their own devices.

### Table 13: Completion Rate of the Tools Course

| Status | Number of Participants |
|---|---|
| Enrolled | 276 |
| Enrolled and started | 170 |
| Enrolled but did not start | 106 |
| Passed* | 66 |

* Learners who completed the full course activities, had a cumulative grade in all quizzes of at least 70%, and submitted the output for the hands-on exercises on the SDMX Constructor, the Fusion Metadata Registry, and the SDMX Converter.
Source: Asian Development Bank estimates based on the evaluation report of the SDMX Tools Course.

Figure 20 shows the gender distribution among people who successfully completed and passed the course. It reveals that 61% of the first cohort of SDMX Tools Course graduates were men, while women accounted for 38%, and 1% elected to not nominate a gender.

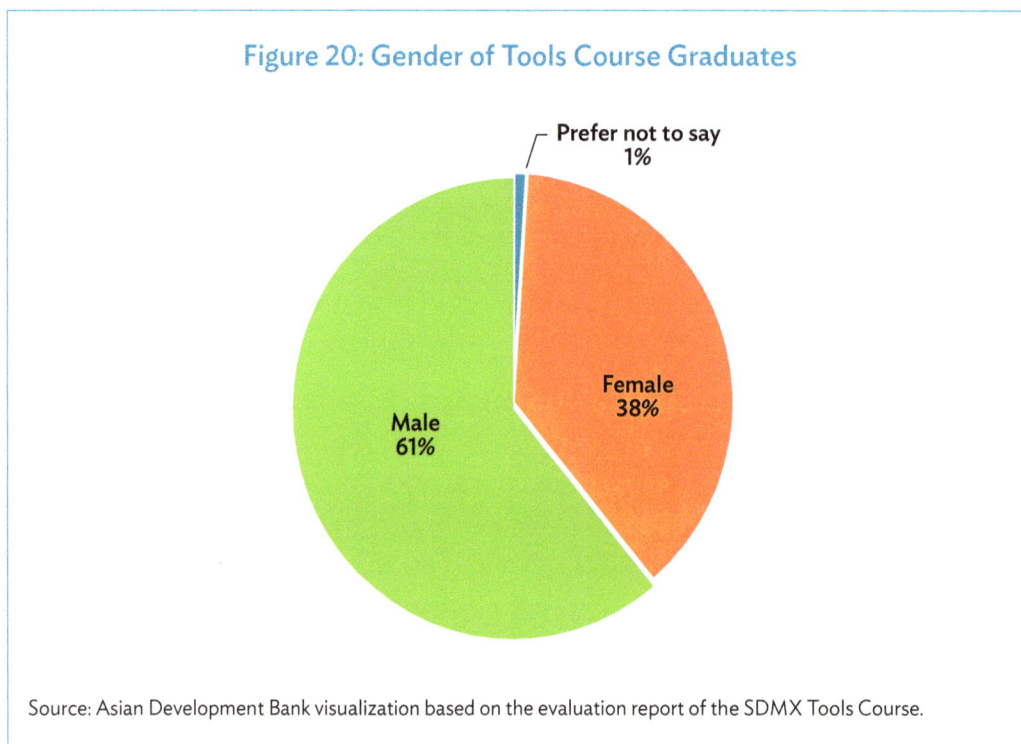

### Figure 20: Gender of Tools Course Graduates

Source: Asian Development Bank visualization based on the evaluation report of the SDMX Tools Course.

Of the participants who passed the first intake of the SDMX Tools Course, 77.4% were below the age of 44 years, with 6.1% of graduates not specifying their age (Table 14).

### Table 14: Age Distribution of Tools Course Graduates

| Age Group (years) | Number of Graduates | % |
|---|---|---|
| 18–24 | 4 | 6.1 |
| 25–34 | 30 | 45.5 |
| 35–44 | 17 | 25.8 |
| 45–54 | 13 | 19.7 |
| 55–64 | 2 | 3.0 |
| Not Specified | 4 | 6.1 |
| Total | 66 | 100 |

Source: Asian Development Bank estimates based on results of the SDMX Tools Course.

Of the successful participants in the first intake of the SDMX Tools Course, 71% were from ADB member economies while 29% were from other economies (Figure 21). Geographic analysis shows that 81% of the people who passed the course were from Southeast Asia, followed by 17% from South Asia and Central and West Asia combined. Among the participants from ADB member economies, staff from five out of ten NSOs that had expressed interest in using SDMX in the ADB economy assessment survey have completed and passed the Tools Course.

### Figure 21: Origins of Tools Course Graduates

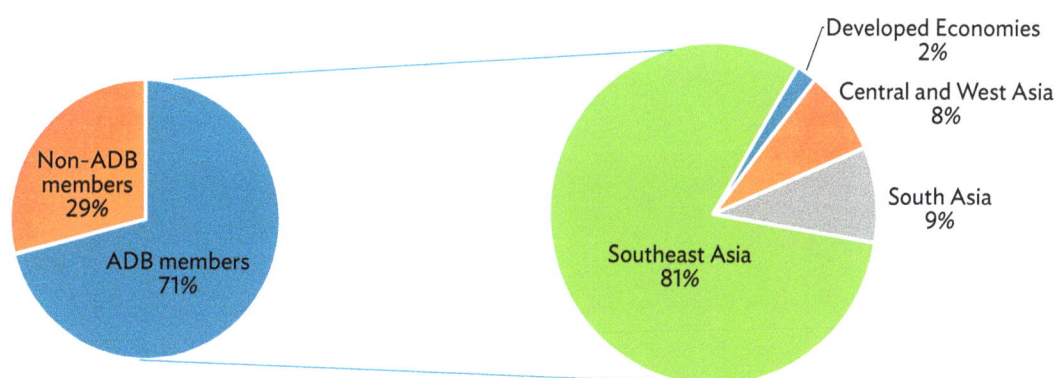

Source: Asian Development Bank visualization based on the evaluation report of the SDMX Tools Course.

Filtering by organizational affiliation, 69.7% of graduates in the first cohort were from NSOs, followed by international or regional organizations and central banks, which together accounted for 24.2% of graduates (Table 15). The least number of learners were from other sectors, with only 1.5%.

### Table 15: Affiliation or Organization of Tools Course Graduates

| Type of Agency | Number of Graduates | % |
|---|---|---|
| National statistics office | 46 | 69.7 |
| International/regional organization | 9 | 13.6 |
| Central bank | 7 | 10.6 |
| Other government agency | 1 | 1.5 |
| Academe | 2 | 3.0 |
| Others | 1 | 1.6 |
| Total | 66 | 100 |

Source: Asian Development Bank estimates based on the evaluation report of the SDMX Tools Course.

In terms of job classification, 56.1% of those who completed and passed the first SDMX Tools Course were statisticians, economists, or held data-related positions. This was followed by IT personnel and people in managerial or supervisorial positions, with a combined representation of 36.4% of all the successful participants (Table 16).

### Table 16: Job Classification of Tools Course Graduates

| Type of Profession | Number of Graduates | % |
|---|---|---|
| Statistician, economist, and data-related job | 37 | 56.1 |
| Information technology-related job | 18 | 27.3 |
| Managerial or supervisorial position | 6 | 9.1 |
| Data dissemination specialist | 3 | 4.5 |
| Others | 2 | 3.0 |
| Total | 66 | 100 |

Source: Asian Development Bank estimates based on the evaluation report of the SDMX Tools Course.

## 9.2.3  Tools Course Evaluation by Graduates

Table 17 highlights the feedback from people who successfully passed the first intake of the SDMX Tools Course. An overwhelming majority of the graduates gave high ratings (*excellent* or *good*) for all specific aspects of the course: (i) clarity of topics, 94%; (ii) presentation style or delivery of the course lecturer, 95.5%; and (iii) relevance of the content to their work, 90.9%. Similarly high ratings were given for both Module 1 and Module 2. Notably, the course received no ratings of "poor" in any category.

## Table 17: Graduate Evaluation of the Tools Course

| Evaluation | Number of Graduates | % |
|---|:---:|:---:|
| **Topics were clearly explained** | | |
| Excellent | 43 | 65.2 |
| Good | 19 | 28.8 |
| Satisfactory | 1 | 1.5 |
| Fair | 3 | 4.5 |
| Poor | 0 | 0.0 |
| **Presentation style/delivery of course lecturer** | | |
| Excellent | 43 | 65.2 |
| Good | 20 | 30.3 |
| Satisfactory | 1 | 1.5 |
| Fair | 2 | 3.0 |
| Poor | 0 | 0.0 |
| **Relevance of content to your work** | | |
| Excellent | 31 | 47.0 |
| Good | 29 | 43.9 |
| Satisfactory | 3 | 4.5 |
| Fair | 3 | 4.5 |
| Poor | 0 | 0.0 |
| **Overall Rating of Module 1[a]** | | |
| Excellent | 37 | 56.1 |
| Good | 26 | 39.4 |
| Satisfactory | 0 | 0.0 |
| Fair | 3 | 4.5 |
| Poor | 0 | 0.0 |
| **Overall Rating of Module 2[b]** | | |
| Excellent | 41 | 62.1 |
| Good | 20 | 30.3 |
| Satisfactory | 2 | 3.0 |
| Fair | 3 | 4.5 |
| Poor | 0 | 0.0 |

[a] Module 1 provides a brief introduction to Statistical Data and Metadata eXchange (SDMX); an overview of SDMX technology tools for various applications; and key use cases, including data reporting, data collection, and data dissemination.
[b] Module 2 covers an overview of and detailed discussion of the select SDMX technology tools: SDMX Constructor, Fusion Metadata Registry, and SDMX Converter.
Source: Asian Development Bank estimates based on the evaluation report of the SDMX Tools Course.

Figure 22 shows the detailed completion rate by each SDMX Tools Course activity.

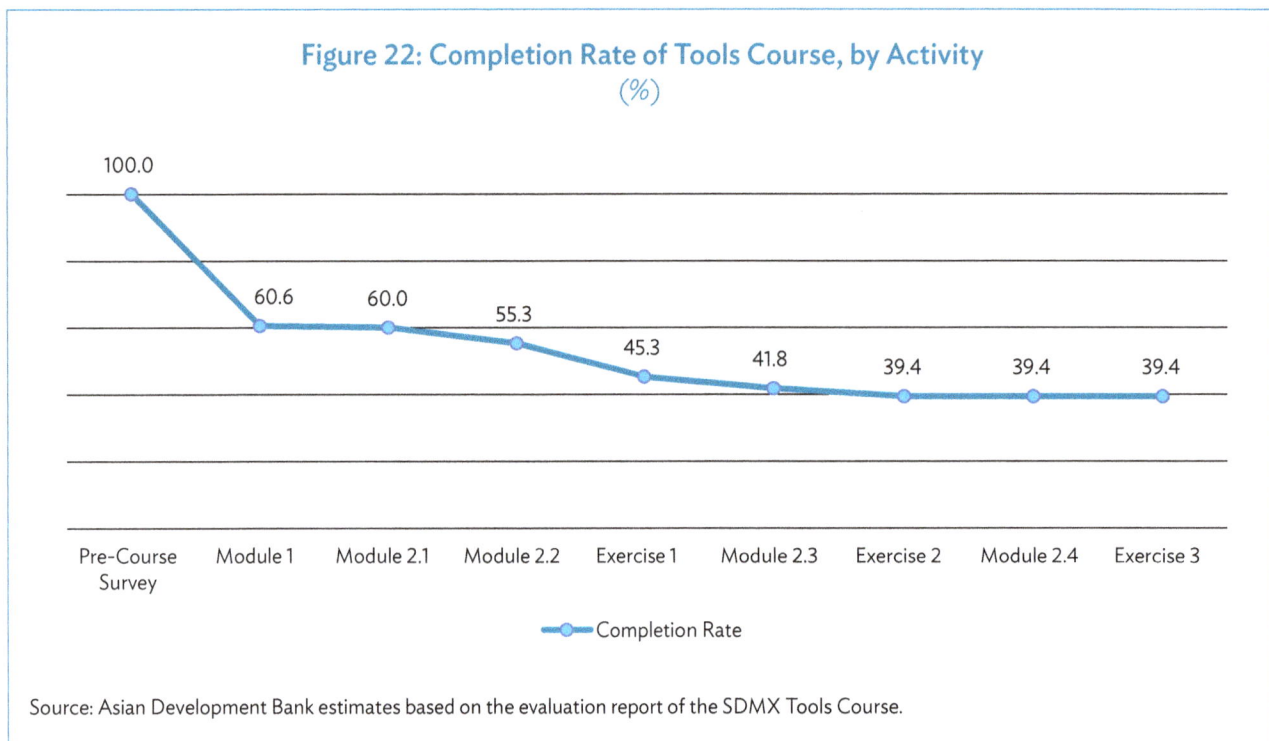

Figure 22: Completion Rate of Tools Course, by Activity
(%)

Source: Asian Development Bank estimates based on the evaluation report of the SDMX Tools Course.

## 9.2.4. Tools Course Results Summary

ADB's 2021 economy assessment survey revealed lack of knowledge on SDMX technology tools as a primary barrier to implementing the SDMX standard. This issue is being addressed by the SDMX Tools Course, which offers in-depth training on tools such as SDMX Constructor, Fusion Metadata Registry, and SDMX Converter.

The course serves as an advanced option for individuals seeking to delve deeper into the implementation of SDMX. Participants must possess a fundamental knowledge of SDMX before taking this course. It is notable that the majority of participants who did not pass the first intake of the SDMX Tools Course had not taken or completed the SDMX Foundation Course.

Most of the first cohort of SDMX Tools Course graduates fell within the working-age demographic and held occupations as statisticians, economists, data-related professionals, IT personnel, or were employed in managerial or supervisory positions. Overall, a strong geographic and gender balance was achieved among the first cohort of graduates.

Important note: The online SDMX Tools Course remains open for enrollments.[15] The course offers remote and cost-efficient learning, with self-pacing to accommodate the participant's preferred schedule.

While enrollment in the SDMX Tools course is open to all, it is important to emphasize that this is an advanced SDMX course requiring participants to have foundational knowledge of SDMX. Course participants and other interested people are encouraged to make use of the SDMX User Forum for any queries, clarifications, or discussions related to SDMX.[16]

---

[15] Access enrollment details via https://elearn.adb.org/course/view.php?id=520.

[16] Access the SDMX User Forum via https://www.yammer.com/unstats/#/home.

# 10 CONCLUSION

ADB, through its technical assistance, has been instrumental in enhancing and modernizing the statistical systems of its developing member countries across Asia and the Pacific.

By providing expert guidance, technical resources, and capacity development on Statistical Data and Metadata eXchange (SDMX), the bank has supported national statistics offices (NSOs) in implementing the SDMX standard within their national statistical systems.

Results from ADB's 2021 SDMX Economy Assessment Survey played a key part in designing and customizing successful SDMX e-learning courses, which were developed in collaboration with three development partners. These courses are tailored to address varying proficiency levels among NSO staff and other participants, from beginners to those with intermediate or advanced knowledge of SDMX. The training covers all aspects of SDMX implementation, from data modeling and structure customization to specialist training on SDMX tools for applications such as data collection, reporting, and dissemination.

This special supplement to *Key Indicators for Asia and the Pacific 2024* is yet another step in ADB's advocacy for progress in the region's statistical systems. It provides a comprehensive assessment of the SDMX standard, illustrating its pivotal role in modernizing and enhancing the efficiency, quality, and accessibility of statistical data exchange across various domains.

Through detailed exploration of key usage applications, benefits, challenges, and real-world implementations, the supplement illustrates the transformative impact of SDMX on global statistical practices. Notably, it highlights the increasing adoption of SDMX by NSOs, and outlines the significant potential to enable innovation in data activities and streamline data operations.

The importance of structural data modeling to support and sustain these improvements has been emphasized, while challenges such as the need for upskilling and demand for open-source tools are acknowledged, along with strategies to overcome these obstacles. The inclusion of SDMX e-learning courses further emphasizes the commitment to building capacity and fostering a community of practice.

This supplement should serve not only as a valuable resource for stakeholders involved in statistical data management but also as a call for the broader adoption and integration of the SDMX standard, ultimately contributing to more informed decision-making and evidence-based policy development. Overall, two main lessons are clear: There is strong regional and global demand to implement SDMX; and online learning is an effective upskilling strategy.

The authors hope the supplement demonstrates the crucial role SDMX plays in standardizing the exchange of data across different stages of the statistical process— including reporting, collection, production, and dissemination—making it more efficient for organizations to share and use statistical information.

# REFERENCES

Ehrmann, H., Tissot B., Başer E., Hülagü T., 2016. *IFC Report 4 Central banks' use of the SDMX Standard. Bank for International Settlements.* https://www.bis.org/ifc/publ/ifc-report-sdmx.pdf.

International Monetary Fund (IMF). Dissemination Standards Bulletin Board. National Summary Data Pages. https://dsbb.imf.org/nsdp (accessed 23 May 2024).

International Organization for Standardization (ISO). 2023. Statistical Data and Metadata eXchange (SDMX). ISO 17369:2013. https://www.iso.org/standard/52500.html.

National Statistical Office of Thailand (TNSO). 2023. TNSO SDMX Documentation. Bangkok.

Paris21. 2021. Data Flow Analysis Framework: Guidelines for Analysing Data Flows in National Statistics Offices. https://www.paris21.org/dfaf.

sdmx.io. Essential SDMX Structural Modelling. https://www.sdmx.io/resources/elearning/essential-sdmx-structural-modelling/.

Statistical Data and Metadata eXchange (SDMX). 2016. SDMX *Content-Oriented Guidelines.* https://sdmx.org/wp-content/uploads/SDMX_COG_2016_Introduction.pdf.

——. 2018. *Guidelines for the Creation and Management of SDMX Code lists.* https://sdmx.org/wp-content/uploads/SDMX_Guidelines_for_CDCL.docx.

——. 2019. *SDMX Implementation Status.* https://sdmx.org/wp-content/uploads/SDMX-implementation-status.pptx.

——. 2020a. *The Business Case for SDMX.* https://sdmx.org/wp-content/uploads/Final_Approved_Version_Business_Case_for_SDMX_Aug2020.pdf.

——. 2020b. Validation and Transformation Language (VTL). https://sdmx.org/?page_id=5096.

——. 2020c. SDMX Glossary. https://sdmx.org/wp-content/uploads/SDMX_Glossary_Version_2_1_December_2020.htm#_Toc59116707.

——. 2023. What is SDMX? https://sdmx.org/?page_id=3425.

Statistical Office of the European Union (Eurostat). SDMX Infospace, SDMX Explained. https://ec.europa.eu/eurostat/web/sdmx-infospace/sdmx-explained#A%20model%20to%20describe%20data%20and%20metadata%20(information%20model.

Tissot, B, 2017. Proceedings of the 61th ISI World Statistics Congress 16-21 July 2017. Marrakech. https://isi-web.org/sites/default/files/import//proceedings/STS069-sdmx-a-key-standard-for-central-banks-statistics.pdf.

———. 2018. *Journal of Mathematics and Statistical Science.* 4 (1). http://www.ss-pub.org/jmss/sdmx-a-key-standard-for-central-banks-statistics/.

United Nations Department of Economic and Social Affairs (UNDESA). 2019. *Introduction to SDMX Data Modelling.* https://www.unescap.org/sites/default/files/Session_4_SDMX_Data_Modeling_%20Intro_UNSD_WS_National_SDG_10-13Sep2019.pdf.

United Nations Statistical Commission (UNSC). 2008. *Report on the 39th Session (26 to 29 February 2008).* https://unstats.un.org/unsd/statcom/doc08/DraftReport-English.pdf.

Ward, D. 2015. *SDMX Starter Kit for National Statistical Agencies.* https://sdmx.org/wp-content/uploads/SDMX_Starter_Kit_Version_23-9-2015.pdf.